THE VANISHING COAST

The Vanishing Coast

Elizabeth Leland

John F. Blair, Publisher
Winston-Salem, North Carolina

PRINTED ON ACID-FREE PAPER

PRINTED AND BOUND BY R. R. DONNELLEY & SONS

DESIGNED BY DEBRA LONG HAMPTON

LIBRARY OF CONGRESS CATALOGING-IN-PUBLICATION DATA

Leland, Elizabeth, 1954–
The vanishing coast / Elizabeth Leland.
p. cm.
ISBN 0-89587-149-1
1. Atlantic Coast Region (S.C.)—Description and travel.
2. Atlantic Coast Region (S.C.)—Social life and customs.
3. Atlantic Coast Region (N.C.)—Description and travel.
4. Atlantic Coast Region (N.C.)—Social life and customs.
5. Leland, Elizabeth, 1954– —Journeys—South Carolina—Atlantic Coast.
6. Leland, Elizabeth, 1954– —Journeys—North Carolina—Atlantic Coast.
I. Title.
F277.A86L45 1992
917.56—dc20 92-6498

TO MY FATHER,

JACK LELAND,

WHO INSTILLED IN ME A LOVE OF THE COAST

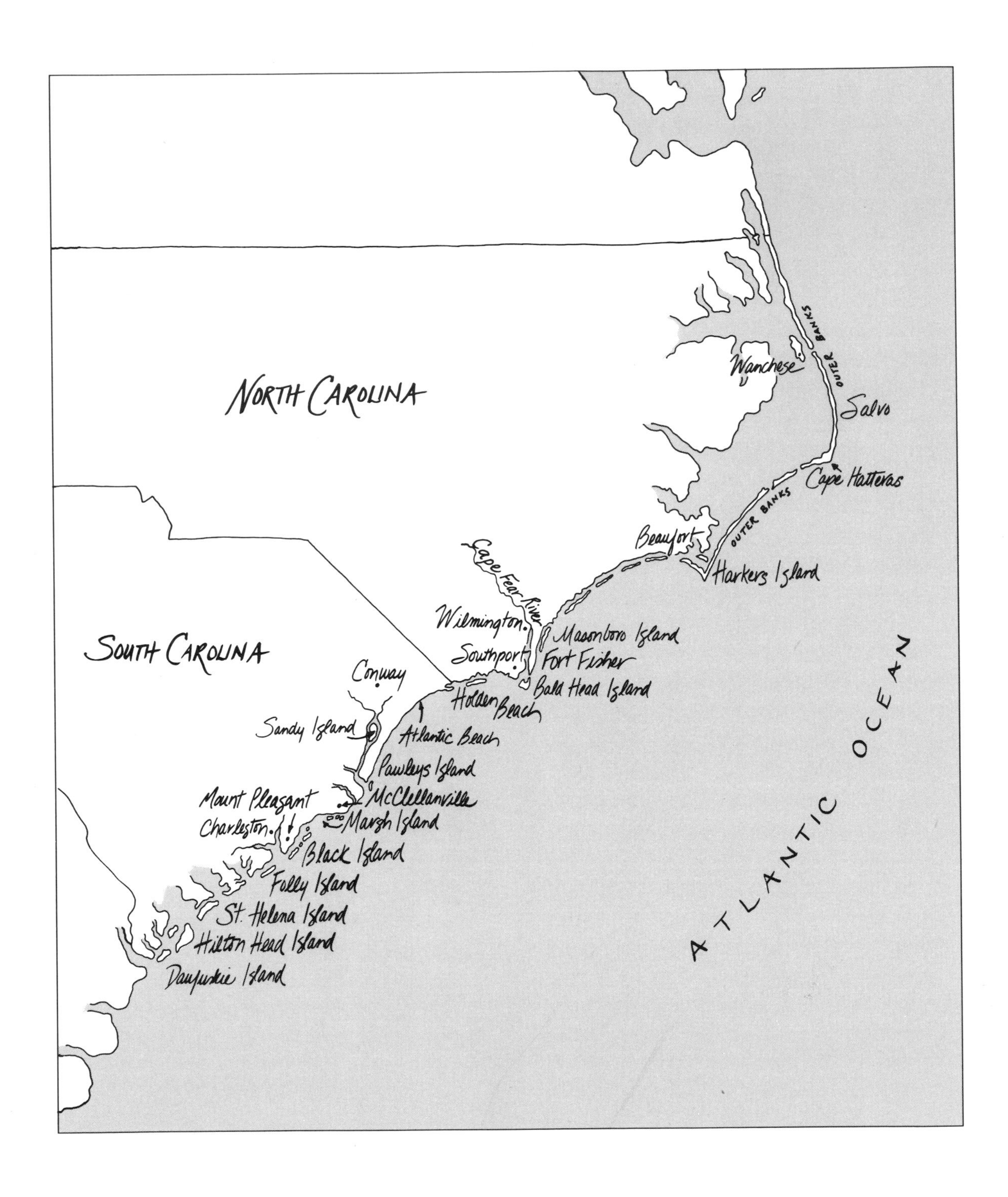
NORTH CAROLINA
SOUTH CAROLINA
ATLANTIC OCEAN
OUTER BANKS
OUTER BANKS
Wanchese
Salvo
Cape Hatteras
Beaufort
Harkers Island
Cape Fear River
Wilmington
Masonboro Island
Southport
Fort Fisher
Bald Head Island
Holden Beach
Conway
Sandy Island
Atlantic Beach
Pawleys Island
Mount Pleasant
McClellanville
Charleston
Marsh Island
Black Island
Folly Island
St. Helena Island
Hilton Head Island
Daufuskie Island

Contents

PREFACE

Deserted beaches stretched for miles when I was a child. Back then, in the 1950s, sea islands like Dewees and Bald Head were home to great white egrets, ospreys, and kingfishers—few people except an occasional fisherman. Kiawah and Hilton Head islands were known for their channel bass and speckled trout, not for their golf and cocktails.

The coast of the Carolinas was extraordinary in its solitude and simplicity. You could walk for hours and meet no one. You could fish all night and fill your freezer. You could skinny-dip in tidal creeks and not get caught.

Those days are vanishing. Drawn by beauty and fueled by profit, we are wrecking our coastline more thoroughly and irreparably than any northeaster could.

Places I knew as a child I no longer recognize. Picture a saltwater creek a few miles up the coast from Charleston, a creek that winds through the marsh in no hurry to get to the ocean. It's narrow enough to jump across at its inland end, but wide enough for a shrimp boat where it empties into the Atlantic near the Isle of Palms. On Saturdays and Sundays, my brothers, sisters, and I would dress in old bathing suits and worn-out tennis shoes and set out from the house, single file across the marsh, a ragtag army of children with cast nets slung over shoulders and bushel baskets in hand.

When we got to the steep, muddy bank, we half-walked, half-slithered into the shallow water. We knew every turn, every drop-off, every oyster bed. The mud was so soft we sank to our knees. We loved it. We called it "mud bogging," and we kept at it until we reached deep water, where another creek fed ours, forming a saltwater pool. Our father christened it "the Mud Hole," and there we would shuck our bathing suits and swim buck-naked under the summer sun. As soon as the tide went out and the water was low enough to trap fish and shrimp in deepwater pools, we would cast our nets. On a good day, we would haul in ten or twenty pounds of creek shrimp, a few mullet, and sometimes a flounder or two.

The sun still casts long, breathtaking shadows across the marsh. Great white egrets still wade in the salty tidal pools. Fiddler crabs still play on the muddy banks. But today a multimillion-dollar golf course winds around the Mud Hole. Fancy wooden bridges for golf carts span the creek. Manicured lawns spread to the banks. An observation platform overlooks the private corner where we used to skinny-dip. Now that islands like Kiawah and Hilton Head are overflowing with fashionable resorts, hungry developers are looking upriver, to out-of-the-way places like the Mud Hole, and to the barrier islands, like Daufuskie and Bald Head, which can be reached only by boat. These days, no place is too remote.

Not too long ago, the developers found Princess Pignatelli's palace. It was a magical spot from my childhood, hidden up a creek off the Wando River northeast of Charleston. It was there that my father told us the story of the wealthy South Carolinian who married an Italian prince and built a mansion with a golf course, a landing strip, and a wharf for her yacht. By the time we came along, a fire had destroyed everything except some of the brickwork, but what remained was still enough to fuel our imaginations.

We would crank the five-horsepower Johnson and navigate our johnboat between sand bars and oyster banks until the ruins came into view. We invented grand tales of the princess and her palace hideaway. In spring, wisteria draped the masonry and filled the air with a scent so strong and sweet it overpowered the musty smell of mud. We watched playful otters slide down creek banks. Once, we came upon a bobcat swimming across the Wando River. We motored along beside him. The brackish water spilled off his sleek back as shorebirds nesting on the river's edge dive-bombed, trying to divert attention from their eggs.

Now, there's an exclusive development where the remains of Princess Pignatelli's palace used to be. There are streets and waterfront homes and golf greens. They call it Dunes West, but there are no dunes.

In the summer of 1988, I decided to search out what was left of the places and people of my childhood. I was working as a staff writer for the *Charlotte Observer*, 177 miles inland in North Carolina's southern Piedmont, but I longed to return to the coast. I grew up on the water, in Charleston; my family settled there in 1678. I may stay away for months, but when I drive the two-lane highway toward home, I always imagine I can smell the salt air long before it's possible. I measure the miles by the first sight of Spanish moss and live oaks, by the warm caress of the Gulf Stream and the sound of the surging surf. I know then that I'm home, and I am happy.

My editors at the *Charlotte Observer*, Mike Weinstein and Jim Walser, encouraged me to roam at will. Hundreds of thousands of people invade the beaches of the Carolinas every year, renewing their love affair with the coast. And for three summers, that's what I did. Most often, I steered clear of the well-traveled routes. I drank copper-colored moonshine on Daufuskie Island, bounced on an old-fashioned joggling board in Charleston, climbed on the weather-beaten limbs of shipwrecks on the Outer Banks. I watched Mary Vanderhorst weave baskets outside Mount Pleasant, Fuzzy Spivey rake for clams off Southport, and blind historian Grayden Paul give a tour of Beaufort, North Carolina, by memory. I visited communities endangered by development and others threatened by nature.

My stories appeared in the *Charlotte Observer* as a series called "Cruisin' 17," in reference to U.S. 17, the coastal highway that runs through the Carolinas and beyond. The stories touched readers in a curiously intimate way. Many called or wrote to thank me. Regardless of whether people live near the coast, it is a place they associate with good times, beauty, and mystery. It is a place that belongs to all of us.

The people I talked with—from Daufuskie Island, near the South Carolina–Georgia border, to Wanchese, near the North Carolina–Virginia border—were raised on the water and schooled in the ways of the tides and the whims of the weather. They are strong of heart and will and heritage. They have persevered in the face of encroaching development, changing times, and the ravages of natural disasters. Some have shared in the prosperity of recent coastal growth. Most haven't.

They are people like Geneva Wiley. Photographer Don Sturkey and I met Wiley as she was hanging clothes on a line outside her home on Daufuskie Island. She was wary of us. She'd seen too many strangers on her island, strangers who wanted to take her land. Wiley was born on Daufuskie, the granddaughter of slaves, and the island is the only home she's ever known. She asks no more than to hold onto her tiny plot of soil and her modest house. But new beachfront resorts are driving property taxes so high that she and other lifelong residents may no longer be able to afford to live on Daufuskie. Hers is a simple dream, but it is one that coastal residents can no longer take for granted.

There's still no thrill like the first scent of the sea breeze as you near the ocean, no better remedy than sand and salt water. Despite the zeal of developers, some places and customs remain almost unchanged: the empty sand bars of Bulls Bay, near Charleston; sunrise on uninhabited Masonboro Island, near Wilmington; the fishermen's brogue on Ocracoke.

This collection of stories celebrates those places and their people. It celebrates the history and the future of our vanishing coast.

Acknowledgments

Without Mike Weinstein and Jim Walser of the *Charlotte Observer*, this book wouldn't have been possible. No reporter could find two better editors or two better friends. I am grateful to them and to the *Observer* for giving me the time, the freedom, and the encouragement to write the stories that make up this collection.

The names of photographers who contributed to this project deserve to appear with my name on the front cover. I salute their work and thank them for their camaraderie. I especially wish to thank Don Sturkey, who knew the best places to eat; Jeep Hunter, who knew the best jokes; Bob Leverone for his rendition of "Two Little Girls in Blue"; and Deidra Laird, Gary O'Brien, Jeff Siner, and Jim Gund for their dedication. Thanks also to Roger Mikeal and the countless copyeditors who read after me, and to Andrea Krewson for the idea of a book.

Many people along the coast let us into their homes and their hearts—too many people to name. Thank you. And thanks to the people at John F. Blair, Publisher, for their support and for their vision of this book.

Finally, to Luke Largess who loves the coast as much as I do, may we have calm seas, clear skies, and a steady breeze.

THE VANISHING COAST

Too Young To Understand The Controversy On Daufuskie Island, Sabrena Robinson Plays In Front Of Her Home With A Dragonfly Tied To A String.

PHOTOGRAPH BY DON STURKEY

DAUFUSKIE ISLAND

AN UNCERTAIN FUTURE

RUTHIE KAPUR cut the engine on her 1971 Ford pickup, and the sounds of Daufuskie Island rushed in—the chorus of tree frogs, the caw of crows, the gentle surge of surf in the calm of late afternoon. "Listen," she whispered. "You don't hear no factories. You don't hear no cars. You don't hear nothing but nature and God."

Pines and moss-draped oaks formed a cathedral ceiling over the narrow dirt road. For a moment, Kapur lost herself in the stillness, in recollections of a barefoot childhood played out across Daufuskie's unfenced acres, of "Mullet Hole," where islanders fished for dinner, and of old Haig Point Plantation, where they hunted deer.

There had always been a timeless quality about Daufuskie, an isolated island south of Hilton Head where some residents still do without indoor plumbing. Its salt marshes and dark, silent woods once seemed impenetrable and indestructible.

But time caught up. Daufuskie has changed. On the end of the island closest to Hilton Head, gates and fences now cordon off exclusive developments. Two planks still jut out over Mullet Hole, but islanders can no longer go there. Residents of the new Haig Point development cross the creek on a nearby boardwalk wide enough for golf carts.

For islanders like Ruthie Kapur, development has been bittersweet. Before the developers came, Daufuskie was dying. The boll weevil destroyed the prosperous cotton crop in the early 1900s. Pollution from the Savannah River shut down the oyster houses by 1959. In the years since, the very remoteness that insulated Daufuskie from development—the lack of a bridge to the mainland—had driven young people from the island in their search for jobs.

The Haig Point residential community brought Kapur back. Her husband, Suresh, got a job there as a maintenance supervisor. For the first time in six years of marriage, they could make a living on the island. "I'm against it, and I'm for it," Ruthie Kapur said of development. "I'm against it because I hate to see change. But I'm for it because it's let me live here."

Daufuskie was one of the last untouched sea islands off the coast of the Carolinas. It is barely a mile southwest of Hilton Head across Calibogue Sound, and within view of the skyline of Savannah, Georgia, farther down the coast. The only access is by boat. Because of that, development was slow to come.

Both Haig Point and Melrose Golf Club were designed with an eye to preserving the island's hundred-year-old oaks and magnolias, its Indian oyster mounds, and the ruins of its former slave cabins. But islanders worry that wealthy, predominantly white vacationers will take over the island as development spreads, forcing out the few dozen blacks and even fewer whites who live quietly and simply on Daufuskie's northern and western shores.

There, Daufuskie remains as wild and mysterious as when novelist Pat Conroy taught in the island's two-room schoolhouse in 1969. Roads are often little more than sandy ruts cut through dense woods. Deer wander through the dimly lit interior, while blue and white herons wade through the surf. Conroy chronicled his experiences on Daufuskie in *The Water Is Wide*, which was later made into the popular movie *Conrack*: "There is something eternal and indestructible about the tide-eroded shores and the dark, threatening silences of the swamps in the heart of the island . . . beautiful because man has not yet had time to destroy this beauty."

In 1987, Daufuskie had sixty-one permanent residents. Beaufort County planners estimate that by the year 2000 there will be more than ten thousand. Some longtime residents blame Pat Conroy for drawing attention to Daufuskie with his haunting descriptions of an island so isolated that its children didn't know the name of the Atlantic Ocean.

John "Cleve" Bryan believes development was inevitable. Bryan's grandmother was born a slave on one of Daufuskie Island's cotton plantations. His mother worked in the oyster houses that once lined the Intracoastal Waterway and provided a steady means of support for the mostly black islanders. Bryan crabbed for a few years after dropping out of school at age thirteen, but he left in 1942 at the age of seventeen to find better work, first in Savannah and then for twenty years in New York. He returned in 1987.

"I like records, and in New York I would listen to that song, 'Dixie Road,'" Bryan said. "I'd think of the island. This is the only place I love. I like the peace and quietness. But you can't stop change. This will become another Hilton Head. I see it."

Bryan is a large, muscular man with a strong face. When he first returned to Daufuskie Island and set to work clearing his five acres of land, his home was a makeshift camp with a blue tent, a red pump, and a shaving mirror hung from a pine tree. Every afternoon when tourists from Hilton Head came ashore for bus tours down Daufuskie's dirt roads, they stopped and stared. "They all want to take your picture," Bryan said. "There's always one joker in the back jumping like a jumping jack."

Jackie Robinson worries islanders will be forced off Daufuskie. "It's going to be a rich man's paradise," he says.

Photograph by Don Sturkey

TOURISTS INVADE DAUFUSKIE ON BUSES
TO SEE HOW ISLANDERS LIVE.

PHOTOGRAPH BY DON STURKEY

Islanders have been so inundated by media and tourists that few welcome strangers. They value privacy, but find themselves with less and less. When Daufuskie resident Bertha Stafford hangs clothes to dry, she also hangs a hand-scrawled sign warning "Wet Clothes," an admonition to the construction trucks that kick up dirt along the dusty roads. The trucks, ferried across the sound on barges, are lumbering reminders of the changing face of Daufuskie.

Islanders can no longer roam at will, and neither can their animals. Developers didn't like it when straying cows ate grass on the golf courses, so they insisted that cows and goats be tethered. There's even been talk of requiring the few people on the island who own cars to get driver's licenses and automobile insurance.

"The island is gone as far as we knew it," said Billie Burn, an island historian. "Something had to happen, though. Either the old were dying or the young were leaving because there was nothing here for them. But there's no such thing as a little development. Eventually, when all of us older people die out, they will throw so much money in front of the heirs that they can't turn it down. This entire island will be for the elite, and not one person who was originally here will be here."

Developers say that's not their intention. "They were here first," said Jeff Quinn of Melrose Golf Club. "We're not about to drive them off the land. We've tried to make the change as easy as possible. We know that if you turn on Daufuskie and its people, you're going to get burned."

But Haig Point's Jack Barry conceded that "eventually most of the island will be developed." Plans call for condominiums along the Intracoastal Waterway, and more golf courses and houses.

The large, undeveloped areas on the island used to be cotton plantations, which flourished until the boll weevil struck. Melrose Plantation was the best known of Daufuskie Island's seven plantations. It was built in 1848 by descendants of Captain David John Mongin, Daufuskie's first white owner. King George II of England had granted ownership of Daufuskie—recorded as D'awfoskee in early days—to Mongin in 1740. By then, most of the island's original settlers—Cusabo and Yemassee Indians—had been killed or driven off.

Melrose Mansion overlooked the Atlantic Ocean. It had five acres of rose gardens and was known throughout the Low Country for its elegant parties. The mansion burned in 1912, but its bricks still littered the beach when the Melrose Company arrived on Easter 1984 and decided to build a country club.

Construction began in October 1985. The resort was designed to include a 52-room inn "bigger than the White House," 200 beach cottages, 350 half-acre lots, and an 18-hole oceanfront golf course designed by Jack Nicklaus. It is the kind of resort, islander Billie Burn says, where people are likely to indulge in champagne and caviar, not the sardines and saltines sold in the island store. Melrose President Bob Kolb called it "the crème de la crème of national country clubs."

Nearby Haig Point was advertised as a "residential retreat." Situated on the island's north-

Because Of Development On Daufuskie,
Animals Must Now Be Tied Up Or Fenced In.
David Robinson Moves The Family Goats.

PHOTOGRAPH BY DON STURKEY

Golfers Play Where Planters Once Grew Cotton On Daufuskie.

Photograph by Don Sturkey

ernmost tip, it commands a sweeping view of Calibogue Sound. Developers divided the plantation into 950 lots. They barged a 1910 mansion up the Intracoastal Waterway from St. Simons Island, Georgia, restored an 1873 lighthouse for use as a two-bedroom inn, and cleaned up the ruins of some 200-year-old slave quarters. The lawns at Haig Point are lush green and well-manicured. Street signs are meant mostly for golf carts, since there are few cars on the island. They bear names like Night Heron Court, Magnolia Court, and River Marsh Run.

On the other end of Daufuskie Island, the undeveloped end, there are no street signs. Islanders have never needed any. As Ruthie Kapur put it, "This road is from here to there and that road is from yonder to thither. That's all we need to know."

For the islanders who have lived off the land and the sea for many years, Daufuskie is much more than a scenic resort. It has been home to their families for generations, many directly descended from the slaves who worked the cotton plantations. Their few acres are hard-won and dearly loved.

Beachfront resorts have driven property taxes so high that islanders like Bertha Stafford worry that they won't be able to afford to live on Daufuskie. Stafford owns two acres and a small wooden house that some people call a shack, but she calls home. Her property was taxed at about $80 in 1982. By 1990, the tax was $390. "I been here sixty-five years," she said, "and we ain't gone around changing our place. We don't bother them people. And these people shouldn't be bothering us and making our taxes go up because of their development. Let them stay on their side. Raise their taxes on that side. I'll stay on mine."

Geneva Wiley lives on the small plot of land she and her husband, Richmond, bought when they married in 1920. She was born on Daufuskie, the granddaughter of slaves. She is slightly bent, favoring a knee arthritic from long hours of standing and shucking oysters at the factory houses. She speaks in the soft cadence of Gullah, the language of many sea-island blacks. "I donna wa fa dey wanna be so hasty wid developing," she said. "When I pass over, I wan my chillun to have mine."

Charlie Ward, Ruthie Kapur's father, feels the same way. He came to Daufuskie about 1926 with his parents, moving down the coast from the Awendaw community, northeast of Charleston. Ward was eight years old then. He attended the island's one-room schoolhouse for whites, which now houses the post office. Ward left Daufuskie in 1934 at age sixteen. He worked at a Savannah paper mill before retiring to an island bluff overlooking the waterway.

He opposes development, but like other islanders admits, "I'm not disappointed with the way it's being done." Ward owns eight acres, part of the original fifty-three acres his father bought for $1,600 in the mid-1920s. Investors have offered as much as $1.5 million for Ward's portion of the land. "I say, 'It's not for sale.' They keep telling me that anything is for sale for the right price. It's not. This is me. Hilton Head was just like this island before they developed there. This eight acres will never sell and never change as long as I'm alive."

Alana Robinson Rides Her Bike Outside Her Grandmother's House On A Part Of Daufuskie Untouched By Development.

Photograph by Jeff Siner

When A Fence Went Up By Barry Chaplin's House,
The Message From The Resort Was Clear:
"We Don't Want You In Here."

PHOTOGRAPH BY DEIDRA LAIRD

THE TWO FACES OF HILTON HEAD

FOR BARRY CHAPLIN, part of the island died the day the fence went up.

Seven feet tall and topped by barbed wire, it forms a garish backdrop to the community where several generations of Chaplins have lived. The Chaplins are one of hundreds of black families who have called Hilton Head Island home since the 1800s. For most of those years, they had the run of the island. Until the mid-1950s, only two white families lived on Hilton Head.

But when Hilton Head Plantation built the fence separating its exclusive development from the Chaplins' modest brick home, the message was unmistakable for fourteen-year-old Barry: "We don't want you in here."

After thirty years of fancy resorts and golf courses, many of Hilton Head's black families have been pushed aside. Some have lost their land to soaring property taxes. Others who've held on now find themselves cordoned off like the Chaplins, with fences and guard gates and No Trespassing signs.

"It feels terrible," said Emory Campbell. "When I grew up, we could go anywhere we wanted to. And we had very few people on the island. Now, to see thousands of people and be restricted, it's a heck of a change."

When a relative dies, Campbell's family must get a pass to go into the old family cemetery, on the banks of Calibogue Sound. The gravestones, many of them hand-carved and some dating as far back as 1861, now sit beneath a multistory condominium complex in sprawling Sea Pines Plantation. The only way you can get into Sea Pines is to live there, pay three dollars per car, or get a special pass.

Nowhere is the tension between old and new more evident than at Virginia Bennett's home. Bennett was born on Hilton Head Island near the turn of the century, the second of eleven children. Her large, wrinkled hands tell of years of hard work shucking oysters in a cannery. The little wooden house built by her husband, Henry, stands in the shade of sprawling live oaks on a rutted dirt road. Time has taken its toll on the sagging porch and weathered siding, as it has on Bennett.

"When I first moved here, ain't no neighbors," she said. Now, a three-story moss-gray resort called The Spa on Port Royal Sound towers outside her back door, making her house look older and smaller than it already is. A matching wooden fence and guard gate separate the resort from the surrounding community of unpretentious homes and trailers.

Bennett is matter-of-fact about the changes. "There's a time for all things, but that time went," she said. "All the motels back there, that's white people. They build what they want. That's their business."

Walter Mack has made it his business. He hates what's happened to the sea islands. Mack works at the Penn Center, a nonprofit community organization on nearby St. Helena Island. He traces his roots in the area to slave days, and he now spends most of his time helping blacks hold onto their land.

"Hilton Head, to me, is a lost cause," Mack said. "All we can do for Hilton Head is look at it and try to let that serve as an example of what not to do. . . . It's a concrete jungle. The way they're building hotels and resorts on the beach, you can't get to the beach because you'll be trespassing on somebody's land, or so they say. I don't want to see that happen to the other sea islands. It's more than just the land. It's an entire culture that's being messed up."

Hilton Head is one of a string of sea islands along the South Carolina coast that was populated mostly by blacks until the surge of coastal development that began in the late 1950s. The island was named for Captain William Hilton, the English sailor who arrived in 1663 to search for new land on behalf of Barbados planters. By 1860, twenty-four cotton and indigo plantations divided the island. But the Civil War and the boll weevil destroyed the plantations, and Hilton Head was left mostly to blacks for the next ninety years. They spoke Gullah. They crabbed, fished, caught shrimp, and raised crops and chickens. They had the run of the island.

By the 1950s, about a thousand blacks lived on Hilton Head. Whites owned most of the land, but only a couple of white families lived there. Occasionally, white landowners hunted on their tracts; others cut timber. In 1952, developers bought most of the twelve-mile-long island. Workers completed a toll bridge connecting Hilton Head to the mainland in 1956, and Charles Fraser began to build Sea Pines Plantation, forever changing the face of what was once called "the Rip Van Winkle of the sea islands."

Development now strains the roads and the imagination. Every summer, speculators knock on the Chaplins' door, asking to buy the family's land. "They say, 'I'll give you a lot of money and build you another house,'" Barry Chaplin said. "We tell them no."

Not everyone can say that. According to Walter Mack, many blacks have lost their land to rising property taxes. Property values on

Hilton Head average fifteen times what they were fifteen years ago. By 1990, the island had 20,000 residents, most of whom were white, and 1.4 million visitors every summer. Island property was valued at $232 million, up from $14 million in 1973.

"I know a man who lost his home for thirty-three dollars in back taxes," Mack said. "Development has been a hardship on a lot of people, especially older people who live on a fixed income. People hold the land real dear around here. The only thing that they've known is the fish and the farm, and through the land they have a sense of pride and self-sufficiency. Once you take that away from them, they have nothing."

Johnny White Walks In One Of The Last Undeveloped Corners Of Hilton Head Island.

PHOTOGRAPH BY DEIDRA LAIRD

Sephus Taylor Expertly Skirts Sand Bars And Oyster Beds.

PHOTOGRAPH BY JEFF SINER

ST. HELENA ISLAND

. . .

THE OYSTERMAN

THE SUN DIPS LOW to the horizon, a steady breeze chills the salt air, and still Sephus Taylor harvests oysters on the solitary riverbank. Legs planted, back bent, he selects one cluster from the thousands in the mud. He splits it apart with three hits from a crowbar, then tosses the largest oyster into a plastic bucket and lets the others fall to the bank. Without looking up, he turns for another cluster, hoping to fill one last bushel basket before dusk and the rising river force him to shore.

His temples gray and his fingers calloused, Taylor still lives by the rise and fall of the tides. He harvests oysters by hand. Winter is his season, when hundreds of fishermen in the Carolinas comb the coastal creeks for oysters.

"I don't believe I want no otherest job," he says. "I been in the river . . . most of my days. Ain't nobody going to get me out now. I just like to be on the water. Nobody bothers you, I know that."

Every day except Sunday, Taylor waits for the tide to fall and expose the mud flats where oysters grow. If the weather and demand for oysters are good, he puts his flat-bottomed boat into a creek off St. Helena Island, a few miles up the coast from Hilton Head. With a hand on his eighteen-horsepower Evinrude motor and an eye to the shore, he winds his way through the coastal tributaries, expertly skirting the sand bars and oyster beds that lurk beneath the surface.

"We call this 'Oyster Field,' that place over there," he says, pointing to an expanse of water. "When the tide goes low, it ain't nothing but bank."

Taylor was raised on the river. The son of a fisherman, he quit school in 1950 at age fourteen to catch crabs and gather oysters. "During them times, things been rough. We have to work to help the rest of them. So my daddy, he tell my teacher to take my name off the books." Taylor never returned to school, educating himself instead in the ways of the river, the

changes of the tide, and the whims of the weather. In the summer, he crabs and farms. In the winter, he harvests oysters.

Taylor's wife, Annie Mae, bore eleven children, seven of whom are still living. But none followed their father into the river. Three work in a New Jersey hospital and three others are in the military, while one lives at home. "These young kids these days, they're not going to do this work," Taylor says. "I believe when we quit picking oysters, nobody else is going to. It ain't that hard. But they ain't going in the river."

The tide flows east toward the Atlantic, carrying with it pieces of marsh grass like so many little boats at sail. Back in his boat, Taylor eyes the bank. He keeps motoring. Ahead, oyster tips peek above the surface. He cuts the engine and drops anchor again.

"Way back in the old days, oysters been plentiful. You could go anywhere and get oysters. You just come on the bank and pick them up. Now, so many things in the river, you don't know what's killing oysters. So many factories right now."

When oysters spawn in summer, they are bitter and milky to the taste. They begin life in the river, the size of a needle point, and later fasten to empty shells on the bank or to each other, growing in clusters at the rate of an inch a year. After Taylor splits the clusters, he leaves behind the smaller oysters to grow another year.

A pelican flies past, skimming the river. An egret wades in the shallows. Taylor doesn't see. His brown eyes are fixed on the bank beneath him, his rubber waders sinking into the soft mud. He thinks only of the next cluster. "You got to work them. You got to beat them out. This kind of work here makes you sleep good at night. I enjoy that life. You work hard for your money."

Two other oyster boats pass on their way to or from different banks. They follow an unwritten code: Don't harvest another person's oysters. "Sometimes you might find a place last a good while," Taylor explains. "You don't want nobody else following."

After the tide turns and begins to cover the bank, Taylor has picked ten bushels. He will sell them for nine dollars a bushel. That's all for the week. The oyster house in Beaufort doesn't need any more. He won't return to the river for a week. The season will keep him busy from winter until May. By then, fishermen in the Carolinas will have harvested 150,000 bushels or more.

There's an old saying that oysters should be eaten only in months with an *r* in their names. Actually, they can be eaten and harvested year-round, but they just taste better in fall and winter, when they're not spawning. The legal harvest season sticks to those months to allow replenishment.

There's another old saying that oysters are aphrodisiacs. Sephus Taylor shakes his head at that one. "That's superstition. If you ain't got nothing in you, you ain't got nothing in you. Oysters not going to do you no good."

When The Remains Of Nineteen Black Soldiers Were Reburied In 1989, It Was A Special Day For George Coblyn, Whose Grandfather Fought With The Fifty-Fourth Massachusetts Regiment.

Photograph by Deidra Laird

FOLLY ISLAND

. . .

An Untold Story of the Civil War

They sailed by the hundreds to Folly Island, south of Charleston—black Civil War soldiers fighting to end slavery.

Some died in battle. But many more were like Frank Newby, a former slave from eastern North Carolina, and Edward Mayhew, a farmer from Massachusetts, who spent grueling hours fortifying Folly Island in hot, humid, and unsanitary conditions. When malaria and typhoid swept the sea island, they died in a lonely military hospital. They were wrapped in blankets, buried, and forgotten.

For more than two hundred years, they lay beneath the sand of Folly Island, part of a chapter of Civil War history that often goes untold and might still be forgotten, if not for the chance discovery of a few Union buttons in a dense wood of palmettos and pines.

In summer 1987, a bulldozer was cutting a rough road through the back of the island for a new development. Civil War buffs Eric Croen and Robert Bohrn earlier had found buttons, bullets, and belt buckles on the island, and they thought the bulldozer might uncover more. "We were walking through there, metal-detecting, and we encountered some signals which turned out to be Union uniform buttons," Croen said. "In the course of digging one of those out, I turned up what I thought was a root. It turned out to be a human femur."

The men called archaeologists. The archaeologists uncovered the bones of nineteen soldiers in unmarked graves in a hospital cemetery. All but one lay on their backs, hands across their abdomens. They were black men, ages sixteen to forty—strong, muscular, with large hands.

While University of South Carolina anthropologist Ted Rathbun studied the bones, other researchers searched for clues about the unknown soldiers. They uncovered a saga of pride and hardship that spanned the coast from Massachusetts to Florida.

Black soldiers weren't allowed to enlist during the first two years of the Civil War. But as the war dragged on and the need for new troops grew, the Union army turned to blacks for help.

Still, only whites were chosen as officers. And while whites were paid thirteen dollars a month plus three dollars for clothing, blacks got only ten dollars a month with three dollars deducted for clothing. Massachusetts blacks refused their pay for eighteen months until the government eliminated the disparity.

More than 175,000 black soldiers fought for the Union army. About 2,000 of them were in the First North Carolina Colored Infantry Regiment and the Fifty-fifth Massachusetts Regiment. The two regiments camped on Folly Island in 1863 and 1864. Some were former slaves, freed after Federal troops took over New Bern in 1862. Others were free black farmers and laborers from Ohio, Pennsylvania, and Massachusetts. They were sent to Folly Island to help in the siege of Charleston, one of the most important Confederate ports, ten miles away.

"It was hot, humid. The living conditions were not good," said Chris Fonvielle, who helped research the regiments. "The drinking water was brackish. And at nighttime, with northerly winds, it was very cold. It was assumed by most military leaders that troops of African descent could better survive the Carolina climates."

They couldn't. During the winter of 1863–64, disease swept the island. The soldiers moved inland, away from the chill ocean breezes, but many died of dysentery, malaria, typhoid. "The whole island was hit pretty hard," said Jim Legg, a Wilmington archaeologist. Of the 402 soldiers from the two regiments who died, 282 died of disease. Some were buried in traditional six-corner pine caskets, others wrapped only in rubberized blankets.

The First North Carolina and Fifty-fifth Massachusetts never became well-known like the first black regiment, the Fifty-fourth Massachusetts, which helped storm Morris Island near Charleston, as portrayed in the movie *Glory*. But the soldiers of the First and Fifty-fifth fought valiantly at Olustee, Florida, and Honey Hill, South Carolina, two Confederate victories.

"Both regiments had distinguished combat careers," Legg said. "They engaged in some really serious combat. They weren't just garrison regiments who didn't do anything and died of disease. . . . They fought like hell."

Private Berry Mitchell of Company B of the Fifty-fifth Massachusetts was shot in the foot at Honey Hill. He returned to Massachusetts and worked as a barber until his death in 1900. His great-grandson, Earl Mitchell of Winthrop, Massachusetts, made a special trip to South Carolina in 1989 for a Memorial Day ceremony honoring the black Civil War soldiers. That day, the remains of the nineteen soldiers found on Folly Island were reburied at Beaufort National Cemetery, just twelve miles from the battleground where Earl Mitchell's great-grandfather fought. "It's like an Indian making a movie and winning," Mitchell said. "It gets your attention."

. . .

A CHILDHOOD PARADISE PRESERVED

THE THICK, DARK WOODS and gold-green marsh of Black Island are reminders of the way the South Carolina coast once looked, and John Ohlandt is determined to keep it that way.

While war raged in Europe in the 1940s, Ohlandt fell in love with Black Island. He would ride the falling tide five miles south from Charleston to the wild island, where he caught spot-tail bass for dinner and swam buck-naked in the creek for saltwater baths. Ohlandt finally got a chance to buy his childhood paradise in the 1970s. Since then, developers have offered him millions to part with it.

They'll have to spend their millions on other islands; Ohlandt resolved that at least one sea island would remain undeveloped. In March 1991, he signed an easement with the Low Country Land Trust that will prevent anyone from building on Black Island. "I want it completely natural," he says. "We've got enough development. We've got enough people. That's why I bought it. Because I wanted to protect it."

Black Island sits in the shadow of the Morris Island Lighthouse near the entrance to Charleston Harbor. It's actually three finger-shaped islands formed of sand dunes and covered by a tangle of crooked oaks, palmettos, longleaf pine, and scrub—50 acres of high land surrounded by 1,950 acres of marsh. The island hasn't changed much since the days when Indians ate oysters off the banks, or since the Civil War, when Federal soldiers chopped pines to build Battery Wagner on Morris Island, from which they bombarded Charleston with their powerful Swamp Angel cannon.

It hasn't changed much since Ohlandt got permission to camp there as a boy. "I would catch the first of the ebb at the top of the tide, and drift," he says. "If it was windy, the rowing was bad and the boat would fill with water. Once you hit Clark Sound, the tide would be fighting you. You had to row like hell."

Rowing from Charleston's old yacht basin

near Calhoun Street to Black Island took as long as five hours. Once ashore, Ohlandt explored the island with Old Black, his coon dog. There was Merritt's Point, where a previous owner shucked oysters to sell to boaters on Lighthouse Creek and gradually built a tiny island of bleached shells. There were freshwater slews that Ohlandt cleared for thirsty rabbits and squirrels. And there were ospreys feeding their young, hawks chasing horned owls, and marsh hens skittering noisily across the creeks.

"In those days, you couldn't get sleeping bags. We tied four to five croaker sacks together, slit the bottoms, and used them as a sleeping bag."

Black Island, which takes its name from an early owner, changed hands several times over the years. Each time, Ohlandt got permission to camp. Then the chance came in the early 1970s to buy the island and protect it.

Ohlandt, a solid-waste consultant with the South Carolina Department of Health and Environmental Control, has made few changes. He cut a four-foot path the length of the island, built a shack at the center, and that was it. Hurricane Hugo ruined much of his work in September 1989, littering his path with pines and crumbling his shelter. There's still no electricity or running water. And there's still only one way to get to Black Island—by boat. At high tide, the trip takes a few minutes from Ohlandt's home on James Island.

That's the way he likes it, but others have grander visions. "You'd be surprised at what people say they want to do with this island," he says. "Let me tell you the one that really just rolled my head. I got a call from people who said they would buy it from me for X million dollars and their plan would leave the island pristine. I said, 'Show me. Draw me up something.' Here's what they wanted to do. They were going to knock every damn tree down and build a golf course. They were going to build a helicopter pad and helicopter people in. They said it would be the most exclusive golf course in the world. They said it would be natural and pristine, all grass. I said, 'Thanks, but no thanks.'"

Ohlandt figured he would always be able to protect Black Island. Then he learned about a nearby piece of coastal property that three sisters named Dill had owned. Even though the sisters stipulated in their will that the land be used as a "recreational sanctuary," developers bought it. Ohlandt didn't want that to happen to Black Island. He acted quickly, deciding upon a legal protection called a conservation easement.

Ohlandt gave up some of his rights—to subdivide the island, develop it, and harvest its timber—to the Low Country Open Land Trust, a private, nonprofit conservation group formed to preserve the vanishing coast. Subsequent owners of Black Island will be bound by the same agreement. The value of the island dropped. But Ohlandt may get a tax break. And most important to him, Black Island will never be developed.

John Ohlandt Cleared This Slew On Black Island
So Rabbits And Raccoons Would Have Fresh Water To Drink.

PHOTOGRAPH BY ELIZABETH LELAND

Elizabeth Leland (center) plays with her brothers and sisters on a joggling board while growing up in Charleston.

JOGGLING BOARDS: A PIECE OF CHARLESTON'S PAST

IN MY HOMETOWN, there's no better place to sit at summer's end than on a joggling board as a sea breeze washes away the mosquitoes and the heat. Joggling boards were as much a part of my childhood as white rice and shrimp creole.

As my mother used to say, a joggling board is to a piazza what mint is to a julep. If you have to ask what a piazza is—they're known as porches outside Charleston—then it's unlikely you know what a joggling board is, either.

It's pretty simple, really. A joggling board is a piece of Charleston history that measures up to 22 feet long and 13 inches wide, a supple board of pine usually supported between wooden rollers 2 feet off the ground. It looks like a giant tongue depressor and acts like a trampoline. There's enough give in the board so that it will bend without breaking or making much noise.

As children, we jumped up and down on joggling boards, trying to bounce high enough to touch the ceiling. Grownups would shoo us away. They preferred to sit and quietly bounce, as gentle as a rocking chair.

It's nice to know that old-fashioned Charleston joggling boards have made a comeback. The man chiefly responsible is Thomas E. Thornhill. Thornhill worked for Charleston Oil Company when I was growing up in the 1950s and 1960s. He retired as vice-president and treasurer in the early 1980s and now sells commercial real estate and joggling boards. As I poked about the city asking about joggling boards, his name came up again and again.

Thornhill has long been interested in woodworking and in Charleston history. He's a former president of the Historic Charleston Foundation. But it wasn't until 1959, when he and his wife, Mardelle, bought a house on Tradd Street in the city's historic downtown, that he became seriously interested in joggling boards. His new home had a long piazza that was ideal for a joggling board, but he couldn't find anyone who made them. So Thornhill measured heirloom boards, studied their construction, and designed his own.

A friend liked the result, so Thornhill made a second board. Then another friend asked for one. It wasn't long before Thornhill's basement was transformed into a workshop and joggling boards began a return engagement on Charleston piazzas.

Thornhill found he couldn't keep up with the demand. "It just got to be too much," he said. "I got a company to cut the parts for me. . . . That got to be too much, too."

In 1970, Thornhill founded The Old Charleston Joggling Board Company with Leonard C. Fulghum, a painting contractor. In its first year, the company made about ten joggling boards. These days, it makes as many as two hundred a year and ships them as far away as Germany.

Thornhill's joggling boards are a lot like the ones I bounced on as a child, with two exceptions—the supports now rest on rockers to give the board more action, and the wood is no longer cut from the heart of longleaf pines. Thornhill says he just can't find longleaf pines big enough. He uses yellow pine instead and has shortened the boards from twenty to sixteen feet. The wood is dip-treated, while the end supports are pressure-treated. Once assembled, the joggling boards are painted Charleston green, the dark black-green of the shutters on many of the city's homes.

Older Charlestonians still talk of courting on joggling boards. It's said that sweethearts would sit on either end and that their bouncing would gradually bring them closer to the center and to each other. As a brochure for Thornhill's company puts it, "Legend has it that there was never an unmarried daughter at the home that had a joggling board."

My introduction to joggling boards came in the late 1950s. My father had one built as a family Christmas present. That same year, one of my brothers got a tool set for Christmas. As we finished opening presents, we heard the sound of a tiny saw at work on the porch. Our joggling board always had a distinctive quarter-inch gash.

What I didn't remember from childhood was that the first joggling board was supposedly made in the early 1800s for a great-great-great-great-great aunt of mine. Charlestonians keep track of family lineages like that. You know all your cousins, as well as your cousins once-removed and cousins twice-removed. And you're taught about your great-great relatives as if they were figures in history books, as indeed they sometimes were.

Mary Esther Kinloch Huger (pronounced Kin-law and U-gee) lived at Acton Plantation in Sumter County, near the community of Stateburg. She suffered from rheumatism. Legend has it that she wrote a letter to cousins in Scotland saying that she had ordered the side of her carriage removed so that a chair could be placed in it to allow her to ride. She told them that riding was her only exercise. The Scottish relatives sent back a model of a joggling board, recommending that sitting and gently bouncing might help her rheumatism. The plantation carpenter promptly built a board according to the model.

And the rest is history.

MARK CHRISTENBERRY SANDS PLANKS FOR JOGGLING BOARDS.
PHOTOGRAPH BY JEEP HUNTER

THOMAS THORNHILL TESTS ONE OF HIS COMPANY'S JOGGLING BOARDS.
PHOTOGRAPH BY JEEP HUNTER

WILLIE DUBERRY WAS ONE OF THE FEW PEOPLE ALIVE IN 1986 ABLE TO TELL WHAT HAPPENED A HUNDRED YEARS EARLIER DURING THE CHARLESTON EARTHQUAKE.

COURTESY OF THE ASSOCIATED PRESS

. . .

THE GREAT CHARLESTON EARTHQUAKE

OUT OF THE GROUND came a roar that shook the farmhouse from side to side and so terrified young Willie Duberry that he ran away in fear of Judgment Day. It was a sultry evening in 1886. The great Charleston earthquake had struck, killing 110 persons and crumbling or cracking nearly every building in the historic port city.

"That was trouble time," Duberry said on the earthquake's hundredth anniversary in 1986. "God showed the people what he could do."

Duberry was one of the few people alive to recount the horror of the night of August 31, 1886. He was a physically feeble but mentally agile 116-year-old. "When it come, it shake up everything," he said. "Colored and white, all of them was scared. They thought it was Judgment. I did, too."

Duberry was living with his parents and twelve brothers and sisters on an eighty-acre farm outside Ridgeville, thirty miles northwest of Charleston. "I come out of the house. All of them stayed in but me. I got in the barn and stayed there until daylight. I had more confidence in the barn than I did the house."

His family survived. His mother's dishes didn't.

The earthquake, estimated in later years at 7.7 on the Richter scale, is the worst ever recorded east of the Mississippi. Tremors shook more than 2 million square miles, toppling chimneys in Atlanta, cracking a skylight 750 miles away in Chicago's Tremont House, and nauseating people from Baltimore to Cincinnati.

It could happen again. Only recently, scientists discovered geologic formations in the sand near Middleton Gardens, north of Charleston, that prove there were at least three major earthquakes before 1886. If a major earthquake struck near Charleston today, scientists estimate it would kill as many as forty-two hundred people and cause billions of dollars in damage. The historic churches and houses in Charleston, already weakened by the 1886 earthquake, are especially susceptible. Even steel earthquake rods added to many of the buildings for stability probably wouldn't help. Most buildings—notably schools and hospitals—aren't designed to resist shock.

Scientists can't predict whether earthquakes will strike Charleston again. Chances of one as severe as the 1886 earthquake are slim. But they say a moderate earthquake likely will strike within the next hundred years.

Unlike West Coast earthquakes, the 1886 earthquake didn't rupture the earth's surface, so scientists aren't sure what caused it. On the days preceding the earthquake, there were several slight shocks, but not enough to worry Charlestonians. Tuesday, August 31, was a typically hot late-summer day. The Charleston *News and Courier* reported that many residents were due back that night from the mountains. Charlestonians recalled a peculiar calm that evening, with no trace of the breeze that usually accompanies the rising tide.

The first dreadful rumble came after dark, at 9:51 P.M. Church bells began to peal. Houses shook. Plaster fell. Bricks flew. "The floors were heaving underfoot," wrote Carl McKinley of the *News and Courier.* "The surrounding walls and partitions visibly swayed to and fro, the crash of falling masses of stone and brick and mortar was heard overhead, and without, the terrible roar filled the ears and seemed to fill the mind and heart, dazing perception, arresting thought."

Young and old—some half-dressed, some injured—rushed shrieking from their homes. Dust from shattered masonry cast an eerie, stifling cloud over the city, and fires burned nearly twenty buildings throughout the peninsula.

"A woman lies prone and motionless on the pavement, with upturned face and outstretched limbs, and the crowd which has now gathered in the street passes her by, none pausing to see whether she is alive or dead," McKinley wrote. "A man in his shirt sleeves, with blood streaming over his clothing from a wound on his head, moves about among the throng without being questioned or greeted; no one knows which way to turn, or where to offer aid."

Survivors cried and prayed aloud at each of seven aftershocks. Most spent that night—and some many more—in streets and parks. Telegraph lines were down, railroads impassable. But as news reached the rest of the world, there came an outpouring of support. Queen Victoria sent her condolences, and cities from Boston to Paris sent money.

Work began quickly, but scars remain. The front of the William Ravenel House at 13 East Battery looks a bit awkward without the Greek portico that was never replaced. The steeple of St. Michael's Episcopal Church still leans a little to one side.

Though Charleston was devastated, the center of the earthquake was about twenty miles north, near where Willie Duberry grew up. The earthquake "shot all to pieces" houses in that rural area, Duberry said, and uprooted the railroad track. "It was all twisted. You couldn't get no train, you couldn't see no whistle blow at that time. I never had seen nothing like that before. . . .

"I never want to see it again."

Earthquake Refugees Camped In Tents In A Charleston Park.
Courtesy of U. S. Geological Survey

Hibernian Hall Suffered Minor Damage In The 1886 Earthquake.
Courtesy of U. S. Geological Survey

The Mosquito Fleet Caught The Fish That Fed The City.

COURTESY OF THE CHARLESTON MUSEUM

THE MOSQUITO FLEET

THE COBBLESTONE STREET is bumpy, and the handlebars shake in the old man's hands. Captain Tom Grant slowly pedals his bicycle past white-columned houses and down narrow streets to his boat on the Cooper River. He is among the last of the Mosquito Fleet sailors.

For more than 150 years, black fishermen like Grant sailed at dawn to the blackfish banks off Charleston. Each evening, city residents were treated to the sight of a fleet of sails peeping over the horizon, riding downwind to harbor. The Mosquito Fleet brought the fish that fed the city. Pushcart vendors ranged through the narrow streets, hawking shrimp and fish and crabs with the familiar cry, "Shrimpee raw, shrimpee raw, fish wid roe and crab wid claw!"

The fleet last sailed in the 1970s as men died, markets changed, and technology intruded. But a few old-timers still tied up on the Cooper River near the spot where their tackle once hung from the sagging ceilings of wooden fishing shacks. And in the summer of 1988, young men, sons of Mosquito Fleet captains, returned with them. They are preserving the memory and the site.

"At the turn of the nineteenth century, it's hard to imagine a fleet of black men plying the waters and making a living when we were in a slave situation," said Melvin Middleton, the leader of the restoration effort. "What people don't remember is there were freed blacks. I think as much about the Mosquito Fleet as I think of Fort Sumter or the Battery. The Mosquito Fleet is a part of history I actually can identify with. It's not something that's just in a photograph."

The origin of the Mosquito Fleet is lost. In 1817, there were reports of freed black men fishing from dugouts off the Isle of Palms. By 1880, you could count fifty sails when the fleet came into harbor. Some of the boats had seven-man crews, men like Tom Grant. In 1917, when he was a boy of twelve, Grant dropped his first hand-line forty miles offshore. Nowadays, he drops his line under a pier and brings up croaker for supper.

But the old days still draw him. "We goes in a group and come back in a group," he said. "That was why they called us the Mosquito Fleet, because we was a fleet of small boats. We all travel together. We looks out for one another."

The boats were flat and narrow, finely drawn at bow and stern, entirely open. They were from eighteen to thirty feet long, with names like *Fear Not, Too Sweet My Love*, and *Swing Low, Sweet Chariot.*

"We used hand-lines," Grant said. "There was no rod or reel. There were four hooks, and if you go out there and if you have four hooks, you have four fish. You could pull them up all day. But you can't. You strain your hand if you do."

On a good day, Grant could catch hundreds of pounds of blackfish, yellowtail, spot, flounder, amberjack, catfish, and porgy. "You don't worry about bait on the blackfish banks. You put a piece of cotton on the lines, put it in, catch a fish, cut that up, and make bait. . . . Locals call them blackfish and sell them for 50 cents a pound. They ship them up north and call them sea bass and sell them for $1.50 a pound."

When it stormed, Grant and his mates would stay ashore for a day of mending nets, caulking hulls, and patching sails. The cypress hulls held up well, but the light canvas sails did not. As the years passed, they took on the look of huge patchwork quilts.

The fleet dwindled after World War I. The boats may have been graceful, but they could not compete in a new commercial environment. When they buried John Green in the early 1970s, they buried the last of the full-time Mosquito Fleet captains. The shoreline where the fleet made its home became an eyesore, littered with bottles, cans, and tires.

The State Ports Authority owns the land, and Melvin Middleton feared it would kick off the remnants of the Mosquito Fleet. Instead, the authority viewed the fleet as a historical treasure. With the authority's blessing, Middleton organized the Mosquito Fleet Association. Its fifteen members—fishermen and their sons—quickly set to work by carting off twenty dumpster loads of trash from the site.

"A couple of guys came practically from the hospital saying, 'I want to be involved,'" Middleton said. "Those older men kept this place so clean. Oh, the pride they had. They used to get the hoes and sweep it clean. . . . You used to walk in here and see a little oasis."

CHARLESTON'S SHORELINE TEEMED WITH THE WORK OF FISHERMEN.
COURTESY OF THE CHARLESTON MUSEUM

CAPTAIN TOM GRANT
PHOTOGRAPH BY JEEP HUNTER

LAWRENCE LUCAS WAITS FOR COLLEAGUE BILL ELLIOTT TO PILOT THE SHIP *ADRIAN MAERSK*, IN BACKGROUND, OUT OF CHARLESTON HARBOR.

PHOTOGRAPH BY DEIDRA LAIRD

Shepherds Of Ships

LAWRENCE LUCAS was born into harbor piloting. For years, you had to be.

In many ports, for many years, harbor pilots belonged to a closed society of hand-picked men. In Charleston, the job passed from father to son. Even now, with state regulation, new pilots must be sponsored by an existing pilot.

Lucas likens it to a fraternity. "It keeps it like a partnership, where everyone knows each other," he says. "You don't want every Joe Blow in it. It's a pretty big responsibility bringing in one of these suckers."

Charleston's fifteen harbor pilots guide every ship going into and out of the harbor. They jump on deck before a ship enters port and when it's about to leave, directing the quarter-master around dangerous shoals and through tricky turns. The pilots know every twist of the channel, the shallows, the sand bars, the flow of the tide. They can set a ship's course dead center beneath a span of the Cooper River Bridge, with only five feet up top to spare.

Lucas was raised on pilots' lore. His father and grandfather piloted ships through Charleston Harbor, passing on a love of the salt water and a knowledge of the channel. His three-year training ended in July 1989. As of fall 1991, he still wasn't a full-fledged pilot. He was waiting until someone retired. It's that competitive.

Pilots make good money—they won't say how good. But they are on call twenty-four hours at a time, seven days at a stretch. They've been doing it for two hundred years, since the days when sails on the horizon signaled the approach of a ship and the first pilot to get there won the job.

As many as twenty ships now pass through Charleston Harbor in a day. They pay the pilots association as much as a thousand dollars for a one-way passage. Every harbor has its pilots. Under federal law, only United States Navy and Coast Guard ships can enter and leave a port without one.

"We've all had some close calls," says pilot Bill Elliott. "But it's a good life."

Elliott figures he sees more beautiful sunrises and sunsets in a year than most people see in a lifetime. He pilots hundreds of tankers, container ships, banana ships—even the navy's sleek, black Polaris submarines. The trip takes him past Charleston's historic skyline, beyond Fort Sumter, where the Civil War began, and twelve miles out into the Atlantic Ocean—then back again. He shares the pride of other pilots in maneuvering huge ships, some nearly as wide as a football field and twice as long, through intricate turns in the channel.

"I've never seen them going under the Cooper River Bridge that they weren't dead center," says Bob Bennett, a retired Coast Guard captain who works with the Charleston pilots. "It's a matter of pride."

One afternoon, Elliott boarded Denmark's *Adrian Maersk* at the Wando Terminal. With Elliott calling directions, the titanic 708-foot-long ship steamed beneath the spans, veering left and right with what looked to be the ease of a twelve-foot sailboat. It sailed past cherry red and lime green buoys, keeping to the center of the channel, where the water is deepest. The ship draws thirty-five feet of water. The channel is only five feet deeper.

Two hours and twelve miles later, Elliott radioed a waiting pilot boat, the *Carolina*. The *Adrian Maersk* was at sea, and the crew could now safely take over. Elliott needed to get off. The fifty-five-foot *Carolina* nudged up to the container ship, their sides bumping in two-foot waves as both vessels kept moving forward at ten knots. Elliott lowered himself down the side of the ship on a metal ladder, as tiny as a window washer on a skyscraper. With a quick step, he jumped from the ladder onto the pilot boat bobbing beneath him. A waiting crew member grabbed his outstretched hand to steady his landing. The *Carolina* veered away sharply, turning back toward Charleston. The container ship set its course for sea.

Getting on and off the huge ships is a dangerous part of the job. In Georgetown's Winyah Bay in 1988, a pilot, Wright Skinner, didn't make the transfer. He slipped as he climbed down an icy ladder from the freighter *Tropical Sky* to the pilot boat. He fell in the bay, and his body has never been recovered.

Charleston's pilots are proud of their record. In the past ten years and forty-five thousand runs, there have been only five accidents in the harbor—groundings and a minor collision, none caused by pilot error.

Over the years, they have come to know each ship's quirks. The pilots are still laughing over one banana ship. The ship eased away from the dock. "Starboard 10!" the pilot called from the deck. The ship kept straight, not veering even 1 degree to the right. "Starboard 20!" the pilot ordered. Still no right turn.

As the ship steamed closer to the shallows in Charleston Harbor, the pilot yelled a final, desperate command: "Hard to starboard!" Then he sprinted toward the control room. A frantic crew member met him halfway. In his hands was the steering wheel.

Pilots also get to know the captains—like the one aboard a Del Monte fruit ship who had a legendary temper. As the story goes, a harbor pilot stood on the ship's deck, calling out orders. No one responded. Puzzled, he went to the control room. The captain, mate, and quartermaster were on the floor, embroiled in a fistfight.

This time, the steering wheel was intact. The pilot grabbed it, ignored the crew, and sailed the ship safely through the harbor.

His Piloting Over, Bill Elliott Must Jump From Ship To Pilot Boat.

PHOTOGRAPH BY DEIDRA LAIRD

Mary Vanderhorst Has Been Weaving Baskets Since Childhood.

Photograph by Don Sturkey

MOUNT PLEASANT

. . .

THE BASKET MAKERS

MARY VANDERHORST sits beside U.S. 17 in the shade of a sprawling oak, practicing a craft that's been handed down for more than three hundred years. She's a Low Country basket weaver, one of about sixty women who fashion sweetgrass into baskets and sell them along the coastal highway.

It's an imperiled Charleston-area tradition.

The coastal development that has made the area popular among tourists now threatens one of the region's more popular tourist attractions. Beachfront construction is destroying the second row of dunes, where the sweetgrass grows. There's so little left that basket makers have been forced to look for sweetgrass as far away as Florida. Construction inland is hurting the basket weavers, too. Some have been forced to move their rough wooden stands from the highway's grassy shoulder, where they have displayed their wares since the early 1900s.

Local historians, builders, and government officials have joined the basket makers in their search for sweetgrass and a safe home along U.S. 17. "These people need to be protected," said Gary Stanton, folk-arts coordinator for the state of South Carolina. "The basket-making tradition is an important part of the whole concept of Charleston. No one wants to see economic conditions or scarce resources adversely affect the continuation of this tradition."

The basket makers themselves may have harvested too heavily in the Mount Pleasant area, across the Cooper River east of Charleston. But their biggest problem is that they can't get to beaches where sweetgrass still grows.

"All the places are built up or private now," Mary Vanderhorst said. "The places we went last year are already posted No Trespassing. Developers have already dug up the grass and put something else there."

After the basket weavers' plight became known in the late 1980s, property owners on Seabrook Island, south of Charleston, opened their shores to sweetgrass harvesting. National-forest officials may open federal land. State officials hope to cultivate sweetgrass.

"People care so much about this," said Dale Rosengarten of the McKissick Museum in Columbia. "First of all, the craft is unusually old. It's maybe the oldest, certainly one of the oldest, African-American traditions in America. It would just be a crime to let it die at this stage."

Most scholars and basket makers believe that slaves brought the craft from the west coast of Africa, and that they were making work baskets for use on rice plantations in the Carolinas and coastal Georgia as early as 1690. As the plantations died, so did basket weaving, until the only place it remained was the Four Mile section of Mount Pleasant.

According to tradition, men gather the sweetgrass and women weave it into baskets. Local women first made baskets for sale in the early 1900s. The baskets became popular in the 1920s, after U.S. 17 was paved from Charleston to Georgetown, drawing more tourists to the area. Today, baskets bring up to two hundred dollars each.

Making the baskets looks easy. It isn't.

Mary Vanderhorst begins by knotting a small bunch of sweetgrass in the middle. She winds the free ends around the knot to make a coil. She then binds the coil with a narrow strip of palm leaf. To this circular base, she adds more and more sweetgrass to make larger and larger coils. Her hands move steadily, with the familiarity born of years of repetition.

Her sister, Liz Fleming, works beside her. "You have to sit down a long time, your butt hurts, you get cramps in your fingers," Fleming said. "It's a dying art. I want the younger generation to learn it, but fewer children are."

An aunt taught the sisters to weave when their hands were scarcely big enough to hold a bunch of sweetgrass. They sell from a wooden stand on U.S. 17 Bypass west of Boone Hall Plantation. In the spring of 1988, Fleming's six-year-old granddaughter sold her first basket—for $13.99.

The sisters use longleaf pine for decoration and bulrush to stretch the sweetgrass. The thicker variety of bulrush, used mainly for work baskets a hundred years ago, is plentiful but hard to manipulate. "If there was no sweetgrass tomorrow, they'd use bulrush," Dale Rosengarten said. "But sweetgrass does a whole different thing. It's smaller and much more pliable. The decorative baskets depend on this fine grass.

"And many of the older women won't switch to bulrush because they've always done it with sweetgrass. It's a tradition in the community, and that in itself is valuable. It's like beauty. The baskets are their own excuse for being. They should be preserved because of that."

Sweetgrass Used To Make Baskets Is Now In Short Supply.

Photograph by Don Sturkey

Young Pelicans Find A Paradise On Marsh Island In Bulls Bay.

Photograph by Don Sturkey

MARSH ISLAND

. . .

Pelican Paradise

THIS IS A BIRD'S PARADISE.

Every spring, thousands of seabirds gather on this sliver of an island to raise their offspring. It's one of the largest nesting colonies of brown pelicans on the East Coast.

Marsh Island is barely visible above the choppy waves of Bulls Bay. The highest point is a pelican's head—4,000 pelicans' heads. By June, the birds have built more than 2,000 nests, and countless thousands of chicks are already hatched and scuttling about. They share the ten-acre island with 7,684 royal terns, 1,744 sandwich terns, 2,776 laughing gulls, 364 tricolored herons, 294 snowy egrets, 10 oystercatchers, 14 great white egrets, 12 glossy ibis, 32 black skimmers, and 38 gull-billed terns. That's more than 17,000 adult birds.

Marsh Island gets its name from the saltwater grass covering its sandy banks. It is part of the Cape Romain National Wildlife Refuge, twenty miles northeast of Charleston as the pelican flies, one of the few remaining places on the coast untouched by developers.

Elegant snowy egrets and tricolored herons strut the shoreline. Brown pelicans with wingspans of eight feet or more float lazily in the muddy surf. Others dive for fish. Thousands sit on thronelike nests of marsh grass, feathers, and flotsam.

At the sound of an approaching motorboat, the nesting pelicans, so graceful in flight, flap their wings in awkward threats against intruders. Hundreds of royal terns, with their distinctive golden beaks, fly away from mottled eggs laid in the sand and, with a noisy chorus, try to divert attention from their unborn chicks. Thousands of wings flutter against the blue sky. The comical cry of the laughing gull pierces the clatter of the terns.

"The pelicans are the first to show up," says biologist Phil Wilkinson. "They begin to nest in April. When they leave in September, this place will be like a ghost town. These particular pelicans will wind up as far south as southern Florida and Cuba. The gulls will go farther. The terns, too."

In 1970, the brown pelican was designated an endangered species.

The birds had died in alarming numbers after the 1940s, when DDT and other pesticides came into widespread use. The chemicals collected in fish eaten by the pelicans and weakened the birds' eggs. The shells became so thin that many broke before hatching.

By 1961, brown pelicans had disappeared from Louisiana and were vanishing in Texas. The problem wasn't as bad in the Carolinas, but by the early 1970s, there were only six hundred pelican nests at Cape Romain.

Since DDT was banned in 1972, the population has steadily increased. Now, fifty thousand brown pelicans nest in Florida, the Carolinas, and the Gulf Coast, roughly the same population as in the 1940s. The brown pelican is no longer an endangered species in the Carolinas.

Marsh Island vies with Bird Key, on the Stono River south of Charleston, as the largest brown-pelican rookery on the East Coast. "The two of them together are the real core of pelican rookeries," Wilkinson says. "They've got more than the entire east coast of Florida."

The island is off-limits to the public.

Tens of thousands of eggs litter the ground. Newly hatched pelicans, looking like prehistoric creatures with purplish black folds of skin, squirm in nests. Older chicks, already grown a foot high in a month, wander boldly across the sandy island. They're covered in snowy down—"just fixing to get their prom dress on," as Wilkinson puts it.

"Two out of three will make it," he says. "By nine weeks, they'll be full-feathered. And then you'll see them running around, flapping their wings for about a week before they actually take off. They feed on menhaden. They will gorge and gain a lot of weight and be heavier than the adults when they fly. The baby fat will hold them over until they learn to eat on their own."

They'll winter further south, leaving in September and returning to Marsh Island in March or April. By their third year, they'll be ready to nest. For the rest of their lives, they'll fly back to Marsh Island every spring to lay their eggs.

A Newborn Pelican Helps Another
By Pecking Off Its Shell.

Photograph by Don Sturkey

HURRICANE HUGO DROPPED HUGE SHRIMP BOATS
INTO PEOPLE'S YARDS.

PHOTOGRAPH BY DON STURKEY

The Night Of The Hurricane

FORESTS TOPPLED. Islands shifted in the sea.

During its terrible journey hundreds of miles across the Carolinas, Hurricane Hugo killed thirty people and rewrote the landscape for generations to come. It ripped off roofs along Charleston's cobblestone alleys, shattered shanties along the rivers, and brought Charlotte to its knees nearly two hundred miles inland. It killed babies in their beds, and it shredded mobile homes from the seaside to the foothills. It stripped the thorns from rosebushes. It blew cotton from the bolls.

Hugo was known to be a killer before it reached the Carolinas. Still, nothing prepared anyone for that night of horror.

At midnight Thursday, September 21, 1989, when Hugo battered the South Carolina coast with 140-mile-per-hour winds, the ceiling fell and seventeen-foot waves broke through the windows at Thomas Williams's home in McClellanville. As muddy seawater forced its way inside the one-story home, Williams and his family ran for the kitchen, where a trapdoor to the attic was located. But waist-deep water had set the refrigerator afloat, blocking their escape.

"Go to the bedroom!" Williams hollered to his wife, Evangeline, and their five children. Once they reached the bedroom, all seven climbed atop a dresser. As muddy water lifted the dresser off the floor, Williams pounded his fist through the plasterboard ceiling. He fell into the water but struggled back to the dresser. Working frantically against the rising water, he lifted his children through the hole and into the attic, where they wriggled into the rafters.

Part of the roof was gone. Wind tore against the family, and rain soaked their clothes as they clutched one another, whispering prayers. The wind roared louder and louder. "Don't scream. You'll scare the children," Williams said into his wife's ear.

"Daddy, stand by me!" their five-year-old pleaded.

They could hear distant screams from the evacuation center at Lincoln High, where the water was rising so rapidly that parents had to hold their children above their heads. Williams lashed his family together with a telephone line. "If they find us," he said to his wife, "at least we won't be scattered all over McClellanville."

For four hours, the Williams family stayed lashed together in the attic, listening to the howl of the storm, comforting one another, and praying. The water got within a foot of their refuge before it stopped. By four o'clock in the morning, it was low enough for them to untie themselves and drop back into the bedroom.

They spent Friday night on a concrete floor in a Georgetown high school.

That Saturday in Charleston, my father and I listened by candlelight as the first grim reports from McClellanville filtered in. We heard the mayor, our cousin Rutledge Leland, pleading for help for his village, isolated and forgotten as the nation's attention focused on Charleston.

At seven o'clock Sunday morning, as curfew lifted, we set off for McClellanville. It was a difficult trip, though not because of the roads—crews had already cleared trees along U.S. 17. It was difficult because of the sadness attached to such a homecoming. My father, Jack Leland, grew up in McClellanville. I grew up on tales of life in "the village."

The closer we got to McClellanville, the worse the damage. Hardly a tree stood in the once-proud Francis Marion National Forest. Homes along the highway were reduced to splinters. As we turned off the highway onto the avenue of oaks leading to the village, we drove through a layer of mud and a maze of chairs and tables and chests tossed from homes by Hugo's fury. Two huge shrimp boats nearly seventy feet long were lodged between trees in the yard of the house where my great-aunt once lived. A beautiful mahogany dining table sat upright fifteen feet above the ground in the branches of an oak tree.

Tom Duke poked his head out of the back of a pontoon boat that was sitting on the shoulder of Live Oak Street. Duke is my father's age—seventy-five at the time of the hurricane. As captain of a shrimp boat, he weathered many storms and hardships in his lifetime, but he spoke with awe as he described Hugo. "The water rose like that," he said, snapping his fingers. "We hollered to everyone to go upstairs. Before we could get to the steps, there was six inches in the house. In five minutes, it was four feet deep."

McClellanville sits on Jeremy Creek, about three miles inland from the Atlantic Ocean. It is a tiny village of moss-draped oaks, narrow streets, and about five hundred residents, the sort of quaint coastal community you rarely find anymore. It began as a summer hideaway for plantation owners in the early 1800s and grew into a seafood center, with many families making their living catching shrimp, oysters, and fish.

As we drove past oaks my father helped plant in 1935, he motioned me toward "the point," the end of the village overlooking the marsh. He has trouble seeing, so he spoke from memory. "There's an old house right here on the end," he told me. All I saw were concrete steps. The house next door had vanished, too.

At every turn, we saw more destruction. Cars

Seventeen-Foot Waves Ripped Through Mary Scott's Home In McClellanville.

PHOTOGRAPH BY JEFF SINER

sat piled up at the high-school evacuation center, one on top of another, three-high in places. Wooden stairs rested in the middle of Pinckney Street. Mud, inches deep, covered the floors of homes. Family heirlooms lay in wet heaps in first-floor rooms wherever the receding tide had dropped them.

My father, a storyteller and retired newspaper columnist, rode silently for much of the way back home to Charleston. At last he spoke, softly and sadly.

"There was a tremendous hurricane in 1822, which sent a tidal surge up the Santee delta, which was one of the greatest rice-growing areas in America," he said. "It destroyed the summering village called Cedar Island at the mouth of the North and South Santee rivers. That hurricane just wiped it out. So the planters who lived on South Santee River began buying lots on Jeremy Creek from Archibald McClellan. The others on the North Santee went to Pawleys Island and built houses. And that's how McClellanville came into being.

"A hurricane built it. A hurricane has now taken it away."

When Her Husband Bought The Cassena Inn In 1972,
Roberta Prioleau Went There Against Her Will.
She Grew To Love It.

PHOTOGRAPH BY JEFF SINER

PAWLEYS ISLAND LOSES A FRIEND

THE CASSENA INN never changed. That's the one thing visitors to Pawleys Island could always count on. The blue paint on the weathered clapboards was always just as chipped as it was the summer before, the screens on the long, narrow porches just as torn. The rusty pipes always leaked. The outdoor shower spout stuck. But as soon as sand was underfoot, the seventy-year-old inn would wrap itself around visitors with a welcome so warm that it brought them back year after year.

To understand the appeal of an inn that owner Roberta Prioleau (pronounced Pray-low) characterized as a "shabby, funny-looking" place, it is necessary to understand something of Pawleys Island.

The four-mile sliver of an island is a throwback to the way beaches were a century ago, when plantation owners went to the coast in summer to escape the disease-carrying mosquitoes of the mainland. The narrow road through the island twists past rambling hundred-year-old homes, many unpainted. Pawleys Island has no high-rise condominiums, no pavilion, no neon signs, no motels. Bumper stickers boast that the island is "arrogantly shabby." Residents look with disdain upon the commercialism of Myrtle Beach, twelve miles up the coast.

"Pawleys Island people are a little bit of a different breed," said Prioleau, the daughter of former South Carolina Governor and United States Senator Burnet Maybank. "They haven't necessarily lived here; they've stayed here every vacation. A lot are extremely wealthy, with extremely high positions. But the thing all have in common is they're all simple. They're just a cheerful, happy sort with a don't-give-a-damn attitude. They come in caked with sand and they don't care."

That's why the Cassena Inn suited them so well. The inn was built by Captain Jack and Elizabeth Brinkley in 1917. Originally called Mrs. Brinkley's Guest House, it consisted of three buildings on the ocean—two single-story structures called "the Boxcars," each with five small rooms, and a two-story building with six

For Rubin Falk Of Asheville, North Carolina, The Ocean Is Just A Short Walk Away From The Cassena Inn.

Photograph by Jeff Siner

WHEN THE SUN WAS TOO HOT, PEOPLE LIKE ROBERT DALTON AND KATHLEEN ROWLAND COULD ALWAYS ENJOY THE HAMMOCK AT THE CASSENA INN.

PHOTOGRAPH BY JEFF SINER

People Like Gerald Merchant Of Chapin
Returned Year After Year To Stay At The Cassena Inn.

PHOTOGRAPH BY JEFF SINER

rooms upstairs and a laundry room downstairs. By the time Prioleau and her husband, Columbia lawyer William "Buddy" Prioleau, bought the property in 1972, a large house had been added inland, across the street. The guest rooms had no televisions or telephones, and the Prioleaus never saw fit to add them.

"We just bought it to hold until the real-estate agent could find someone to buy it," Roberta Prioleau said. "I was forced to come down here, much against my will. And then I fell in love."

Prioleau wintered in Columbia with her husband and summered in the big house on Pawleys Island. There, she served guests old-fashioned Charleston dinners. Her meals of biscuits, corn sticks, salad, vegetables, meat, seafood, dessert, and iced tea soon drew diners from miles around. A reputation and a loyal following were born.

The big house burned down in July 1986, but Prioleau kept the sixteen rooms in the three remaining buildings open. Some of her customers had by then been coming to the inn for so many years that if they beat Prioleau there in the spring, they knew where to find a key. If they happened to leave before she arrived, they simply left a check underneath a mattress.

Prioleau intended to keep the Cassena Inn the way it had always been, with wooden rocking chairs, rope hammock, and linoleum floors that no amount of sand and salt water could harm. She said she wanted it as much for herself as for Pawleys Island and the vacationers who had been coming to the inn for decades.

But circumstances changed. She decided to sell the inn so she could spend more time with her husband. Prioleau vowed to sell only to the sort of person who understood the Pawleys Island way of life. She stipulated that the new buyer couldn't change a thing at the inn, from the worn wooden steps to the concrete floor of the open-air "Paper Plate," where she served fresh fruit and cereal to guests at breakfast time.

"I had about twenty people seriously interested and every single one came back with the same story," she said. "They said they couldn't keep it the same. They used the same expression: 'The numbers don't work.' They said it was because when we bought it we didn't pay as much for it. They said they had to make it larger to make money. The first six I told no. After twenty people said the same thing, I realized it would have to change."

The decision did not come easily to Prioleau, but when she locked the green slatted doors in September 1988, it was for the last time. Developers planned to build a thirty-five-room, million-dollar inn. While the new building would be structurally similar to the Cassena Inn, it would include such niceties as wainscoting and heart-pine floors.

The sale upset islanders as nothing else could. They protested that the new inn would ruin Pawleys Island, that thirty-five rooms were too many. They sued.

In the middle of the legal battle, in September 1989, Hurricane Hugo accomplished what developers had planned. It destroyed most of the inn, and with it a part of Pawleys Island.

Robert Wylie, a pharmacist from Chester, visited the Cassena Inn for the last time just before it closed in 1988. "It's just so comfortable," he said. "There's none of that glitz and all that [like] at Myrtle Beach. You don't have to worry about tracking in the sand. You're right at home here."

Said Vernon Goode of Charlotte, another longtime visitor to the inn, "You lie on those cotton sheets, with an old rusty screen in the window, and the fan is blowing the curtains past the bed. You can hear the ocean. You can hear the crickets. Oh, I just love it. As times go by, some people want to modernize, and they add sheetrock, and carpet on the floor, and pretty soon you've kind of lost the spirit. But they never did that at the Cassena Inn."

Elizabeth Rowland Of Sumter Naps Beneath The Inn.

PHOTOGRAPH BY JEFF SINER

Roberta Prioleau Says She Hated Giving Up The Inn.

Photograph by Jeff Siner

Solomon Pyatt Paddles Toward The Mainland From Sandy Island.

PHOTOGRAPH BY DEIDRA LAIRD

. . .

A Place Apart

THE SUN GLANCES through the cypress trees, and Rose Pyatt shields her tired brown eyes to watch a cabin cruiser pass. "We used to row across the river," she says. "Now, all the boats come this way. I don't trust it."

As the engine fades in the cruiser's wake, an uncommon stillness falls over Sandy Island. Silent shadows dance across Pyatt's gate. A gentle breeze stirs through a forest of oaks and longleaf pines, carrying with it the sweet perfume of wild wisteria. Far away, a woodpecker drills.

For Pyatt, it's another solitary afternoon at home, cut off from South Carolina's Grand Strand by the Intracoastal Waterway. It will be an hour yet until the ferry brings back the island's twenty-six children from public school, and a few hours more before their parents steer their skiffs home from work.

As they've done for more than a hundred years, the people of Sandy Island commute by boat to jobs, school, doctors, and stores. They envy the modern-day conveniences that a bridge would bring, but they jealously guard their solitude, rare along the rapidly developing coast of the Carolinas.

"I would live nowhere else," Pyatt says. At eighty-two, she is the island's oldest resident. She is the fourth of six generations of people to call Sandy Island home.

Before the Civil War, nine wealthy rice planters summered on the island. But it was Pyatt's great-grandfather, a man named Philip Washington, who set up permanent residence. Washington bought a piece of the forty-eight-square-mile island after he was freed from slavery, and he and thirty-one other freed slaves made Sandy Island their home. Even now, residents refer to parts of the island by their plantation names: Grove Hill, Hasell Hill, Holly Hill, Mount Arena, Oak Hampton, Oak Lawn, Pipe Down, Ruinville, Sandy Knowe.

Most of the thirty families on Sandy Island live about a half-mile inland in a cluster of homes called "the village." There, conch shells

THE HERRIOTT FAMILY RIDES HOME FROM CHURCH.

PHOTOGRAPH BY DEIDRA LAIRD

Residents Treasure The Peace And Quiet Of Sandy Island.

Photograph by Deidra Laird

Children Pray At New Bethel Baptist Church On Sandy Island.

Photograph by Deidra Laird

decorate the railings of small houses with brightly colored shutters and tin roofs rusted orange. Rutted dirt roads meander through longleaf pines to the white stucco New Bethel Baptist Church and a grove of abandoned pickup trucks.

Boats are a necessity of life on Sandy Island, but many islanders also own two cars or trucks. They keep one on the island and the other on the mainland. At high tide, it's a five-minute boat ride across the Intracoastal Waterway and down an old rice canal to a dock near U.S. 17 and Brookgreen Gardens. The trip takes ten minutes longer at low tide; avoiding mud banks makes navigation a little trickier. Islanders accept such inconveniences as city dwellers accept traffic and late-night sirens.

Solomon Johnson's arms are muscular from years of rowing, his feet hardened from years of trudging through the deep sand that gives the island its name. "You never get used to it," he says. "The older you get, the harder it is."

Isaac Pyatt, Rose Pyatt's nephew and a Georgetown County deputy sheriff, remembers times when the moon was the brightest light on Sandy Island, and when summers meant floating lazily on the river. "There's always a good neighbor," he says. "But as inconvenient as it is, the people and houses here aren't much different from the mainland."

It wasn't until the late 1960s that telephone and electric lines crossed the waterway. Isaac Pyatt believes that a bridge and development will someday follow. "From Charleston on up to Myrtle Beach, the land is taken up," he says. "They have no place to go but this way. There will be losses with the bridge—motels, condos, restaurants, stores. It'll change a lot. The elderly, they don't want to see certain changes because it's been this way so long. But for the younger generation, it'll be nice."

For years, rumors of a bridge have taunted islanders, who think that easy access to the mainland might help convince their children to come back home after college.

Politicians want bridges to hopscotch Sandy Island from U.S. 17 on the coast to U.S. 701 inland, which would make getting to the Grand Strand easier and getting away from hurricanes quicker. "It's a logical step," says Alfred Schooler of Georgetown. "But that's probably in the distant future."

Schooler, a county-council member for more than thirty years, has mixed emotions about a bridge. "I feel like if the island is developed, these people will be pushed into a corner," he says. "Inevitably when you have development, your taxes go up. You have to pay for amenities. I don't know if it would be a great thing or not."

It's happened before. Longtime residents of some coastal islands find themselves surrounded by barbed-wire fences erected by exclusive resorts. But Sandy Island residents are confident that it won't happen to them. "If I die, that's the only time they'll push me off this island," Rose Pyatt says. "This is home."

Sandy Island Schoolchildren Walk Home From The Dock
Where The Ferry Lets Them Off.

PHOTOGRAPH BY DEIDRA LAIRD

The Wade Hampton Oak Forms A Canopy
Over Conway's Main Street.

PHOTOGRAPH BY DON STURKEY

BUMPER TO BUMPER

THERE'S A SAYING in this coastal South Carolina town: "Eleven million people go to the beach each year. Seven million go through Conway twice."

Even if the figures don't quite add up, the sentiment does. During peak beach weekends, the town of Conway is little more than a traffic jam.

More than a quarter of a million tourists crowd the sixty-mile Grand Strand every day during the summer season. One way or another, it seems that most of them end up driving through Conway—if you can call it driving.

U.S. 501, the main route to the Grand Strand, slows to a standstill entering Conway and all the way to the coast. On Sundays, when beachgoers head home, it can take as long as three hours to creep the fifteen miles from Myrtle Beach back to Conway.

For many tourists, Conway has become a curse word.

For the town's 13,500 residents, tourists are a mixed blessing. Many Conway residents work in Myrtle Beach and depend on tourism for their livelihood. But they don't enjoy the traffic any more than do the tourists. They don't appreciate the reputation Conway has among frustrated motorists, either. "It's putting a bad image on Conway, an image we don't deserve," says Debby Brooks of the Conway Chamber of Commerce. "Conway is one of the nicest towns you'd ever want to see. It's one of the most historic in South Carolina."

A few blocks east of the congested U.S. 501 Bypass, out of earshot of angry car horns, lies Conway's historic section. Plans for the town, originally known as Kingston Township, were drawn up around 1734. The first settlers arrived a few years later. It wasn't until the early 1900s that Conway residents began to build cottages at "New Town," the new summer retreat now known as Myrtle Beach.

In Conway, broad oaks with Spanish moss form a canopy over residential streets where homes date back a hundred years or more. Red geraniums, scarlet sage, and pink impatiens decorate gardens. Hammocks and wooden swings hang from wide porches.

A plaque on a large oak on Main Street commemorates the day in 1876 when Confederate General Wade Hampton brought his campaign for the governorship to Conway. Another plaque, under the shade of an oak on Sixth Avenue, honors soldiers who died defending the Confederacy.

It's ghostly quiet along these streets in the middle of the week. But on weekends, tourists who are disgusted with the traffic on the bypass seek out shortcuts through town that once only the locals knew. Traffic has steadily worsened.

"Of course we love the tourists. We want them to come. But I will be honest with you, it's very difficult," says Lois Eargle, a Conway real-estate agent. "You have to time when you go out with how the traffic will be."

The Exxon service station on U.S. 501 Bypass simply shuts down on Sundays. The station is on the side of the highway leading south, toward the beaches, but on Sundays most of the traffic crawls the opposite way, as vacationers head home. "We don't get any business on Sunday," says Walter Roberts. "We haven't been open on a Sunday in about five years. People don't want to get out of line because they're scared they won't be able to get back in."

Traffic bogs down so badly through Conway that there have been reports of passengers hopping out of cars, dashing into fast-food restaurants to use the toilet, and hopping back into their cars, which have moved forward hardly at all. "Horrendous," Debby Brooks calls it. "It's getting worse and worse, the more the beach resort area comes alive."

That's okay with Rod Gragg, just as long as the traffic moves on through. Gragg is a Conway historian who runs a public-relations agency. "Many of us who live in Conway, like other folks on the Grand Strand, derive our living from the tourist season," Gragg says. "I like them. At the same time, there's a vast difference between Myrtle Beach and Conway. There's a greater sense of history here. People want to see the Grand Strand grow and flourish, but we want to maintain a sense of community in Conway."

Gragg admits that the heavy traffic on U.S. 501 makes it "very uncomfortable for people coming to the beach. Some of us are grateful for that. We don't want them to know what a great town this is. They're liable to turn it into another Myrtle Beach."

Hidden From The Highway That Passes Through Conway
Are Historic Homes Like The Beaty-Spivey House,
Framed By Confederate Jasmine.

PHOTOGRAPH BY DON STURKEY

At Atlantic Beach, Hand-Painted Signs
Have Never Been Replaced By Neon.

PHOTOGRAPH BY DEIDRA LAIRD

ATLANTIC BEACH

THE BLACK PEARL

FOR HARVEY GANTT, there is no other beach.

When Gantt was a child in Charleston in the 1950s, Atlantic Beach sparkled as "the Black Pearl" in the whites-only Grand Strand. Most summers, Christopher and Wilhelmenia Gantt loaded their '51 Ford station wagon with young Harvey, his four sisters, fried chicken, and potato salad and drove north past a hundred miles of beach to the only stretch of the South Carolina coast that welcomed blacks.

They came by the hundreds. Black blues singers spent the night at Atlantic Beach after their Grand Strand gigs; black families crowded the surf on weekends; black newlyweds honeymooned there.

Integration changed all that. Blacks began to forsake Atlantic Beach for flashier resorts like Myrtle Beach and Cherry Grove Beach, only a few fishing piers away but years ahead in comfort. Atlantic Beach lost its luster.

Gantt hopes to restore it. His Charlotte-based architectural firm has drawn up plans for a new Atlantic Beach. "We want it to be an entertainment center for the entire Grand Strand, but also act as a kind of museum of what it was like at one time in our history," said Gantt, a former mayor of Charlotte and candidate for the United States Senate. "It will never be a totally black town anymore."

Atlantic Beach sits between Crescent Beach and Windy Hill Beach, a low-rent district between high-rise hotels. Its four blocks are as faded as the frame homes that line its streets. An eerie quiet hangs over the deserted streets, littered with abandoned cars, empty beer cans, and boarded-up businesses.

The few establishments that operate year-round—places like Zeno's Bar and Brooks Soul Food—are throwbacks to years past, with concrete floors and tattered wooden booths. They offer none of the pizazz that makes the rest of the Grand Strand a $1.4-billion-a-year tourist mecca. There are no curbs in Atlantic Beach, and no gutters. Other beaches have neon signs, but hand-scrawled signs announce the Holiday

High-Rise Development Halts Abruptly At Atlantic Beach.

PHOTOGRAPH BY DEIDRA LAIRD

Motel and the Manhattan Inn at Atlantic Beach.

Longtime resident Earlene Woods has listened to promises like Gantt's before. Those promises turned out to be as empty as Atlantic Beach's motel rooms are most of the year. But she hopes this time will be different. The town has been awarded a $250,000 planning grant through the governor's office. Local officials contracted with a Columbia-based developer, which in turn contracted with Gantt's architectural firm. "We've had our hopes built up and let down in the past," Woods said. "This time, if the governor would give a grant of $250,000, there's hope."

Woods owns a hotel, an apartment building, and seven lots in Atlantic Beach. Like other property owners, she has stubbornly refused to sell over the years, when developers offered six-figure sums. "I felt if it was worth that much to someone else, why couldn't I hold onto it for future generations?" said Woods, who has thirteen great-grandchildren.

Blacks have owned most of the town since the 1930s, when laundromat proprietor George Tyson, a black, bought property from three whites. Neighboring beach towns merged to form North Myrtle Beach in 1968, but Atlantic Beach refused to join. "In the past, developers came in and wanted condominiums and high-rises," Woods said. "We want a nice, quiet, developed town that anybody can come to."

Local leader Joe Montgomery remembers what happened when sea islands like Hilton Head and Daufuskie fell into the hands of developers. He has vowed that the same thing will never happen to Atlantic Beach. Redo the town, Montgomery says, but redo it under black control. And redo it so that the poor will be as welcome as the rich.

Harvey Gantt shares Montgomery's vision. "There's a certain historical aspect of Atlantic Beach that should never be lost," he said. "We want to build bathhouses, so that for those who can't afford hotels, motels, there will always be an Atlantic Beach where they'll be able to go to the beach."

Gantt and others envision a public park and boardwalk along the beach, with hotels located farther back. The main street, Thirtieth Avenue, would be lined with open-air markets featuring ethnic food and Caribbean music—the flavor would be similar to that of the French Quarter in New Orleans. A nonprofit corporation of town residents would control development. Owners of property targeted for development would have the choice of participating in the project by selling or leasing. If they refuse, the town could seize control through its power of eminent domain, according to Joe Montgomery. Montgomery plans to raze his worn-out liquor store and replace it with a welcome center.

Some landowners and residents worry. If development falters, they could lose what they've jealously guarded for generations.

But for others, the risk seems worth it. "I remember when you couldn't walk here because there were so many people in the street," Elmore Davis said as he idled away an afternoon in front of empty Zeno's Bar and looked out over empty Thirtieth Avenue. "It was real fun. We need progress."

Thelton Gore, a commercial fisherman who operates the Atlantic Beach Pavilion, has watched black customers flee to nicer beaches. "People have just about quit coming. We don't have things to offer. . . .

"It's like we're sitting at a card table. We either got to get in the game or we're going out backwards. It's time for a change."

Elmore Davis Looks Forward To Better Days For Atlantic Beach.

PHOTOGRAPH BY DEIDRA LAIRD

FISHERMEN LIKE HARRELL PADEN NO LONGER ARE ALLOWED TO DRIVE PAST THE GUARD GATE TO THE WEST END OF HOLDEN BEACH.

PHOTOGRAPH BY JIM GUND

No Trespassing

HARRELL PADEN shook his head in disgust. "No Trespassing," the sign warned. "You can't come in," a woman told him from the guardhouse. "It's private here."

Paden was no newcomer to Holden Beach—a nine-mile-long island in Brunswick County, North Carolina, midway between Wilmington and Myrtle Beach. He had fished from the secluded west end of the island since 1928, when he was four years old. He was so little then he had to hold the rod between his knees and swing it sideways to cast.

But when he returned one afternoon in June 1985, a guardhouse blocked the road. Only people who owned or rented houses would be allowed in, even though fishermen and swimmers had used the west end of the island for years.

"They're closing the beach to the public," Paden said. "The city picks up garbage down there. Police cruise down there. There's a city easement for a water line that runs right down that road. But they won't let us walk down that road. They want our public money and them a private beach."

The beach still belongs to the public—technically, at least. But in many places, the public can't get there.

Gates block the entrance to Fripp Island in South Carolina and Figure Eight Island in North Carolina. Fences cordon off Hilton Head communities. No Trespassing signs greet visitors to the Outer Banks. "It's only going to get worse," warned Jimmy Chandler, an environmental lawyer in Georgetown, South Carolina. "It doesn't do a whole lot of good to say the public owns the beach, but the only way to get there is to take a boat from the ocean and swim in. We've got to do something."

"There's no question from a development point of view why they do it," said Durham lawyer James Maxwell. "If you can make a place more exclusive, it becomes more valuable. . . . But the public has a right of access to the beaches."

The right of public access dates back to an old English law called the Public Trust Doctrine. The doctrine states that beaches and waterways are owned by the state for public use.

Getting to beaches in the Carolinas was never a problem until recently. Visitors just walked across vacant land, and few owners complained. But since Hurricane Hazel leveled the coastline in 1954, beach construction has been unrelenting. New houses block old footpaths.

Even in places where the state has constructed walkways for public use, there's often not enough parking close by. In 1981, North Carolina ranked last among coastal states in providing public beach access. Since then, the state has spent $2.4 million on parking and dune crossovers. But that's not enough.

"It's become very difficult for the average citizen who wants to go to the beach for the day or for somebody who has a house two or three rows back," said Bill Keese of the North Carolina Division of Coastal Management.

South Carolina beaches face the same problem. Bill Schwartzkopf has watched public

Harrell Paden Examines The Poles He Used To Bring To Holden Beach.

PHOTOGRAPH BY JIM GUND

beaches go virtually private at developments like Litchfield By The Sea, DeBordieu, and Inlet Point. "The only way for the public to get to DeBordieu is to go by boat," said Schwartzkopf. "Or if you're brave enough, you could go to Pawleys Island and swim across the channel. But that channel gets pretty swift down there. If you can't get to them and have nowhere to park, what good is a public beach? It's a concern for the entire state. They're not making any more beaches."

Jim Griffin, the president of Holden Beach Enterprises, defends the right of developers to block roads like the one leading to the west end of Holden Beach. "It's a matter of protecting the property and having a type of subdivision down here that a certain part of the public can appreciate. . . . They're looking for something that's quiet and very restrictive."

In 1961, Holden Beach Realty put up the first of a series of barricades to the island's west end. Most were ignored or removed until 1985, when Holden Beach Enterprises bought the property and built a guardhouse. "One day, there were over two hundred cars down there," Griffin said. "We're trying to protect our property. We do not restrict anyone from the strand and we don't intend to. But that road belongs to property owners."

A group of Brunswick County taxpayers sued in 1986, contending that the half-mile road is public. They based their argument on the legal notion of proscriptive easement, which says that a road becomes public after continuous and uninterrupted use over the course of twenty years. The state of North Carolina joined the suit in 1987. "They've effectively created a private beach," Allen Jernigan, an assistant attorney general, argued. "It greatly increases the value of their property to the detriment of the public."

In November 1987, Superior Court Judge Bruce Briggs sided with the developers, who contended that barricades had prevented twenty years of continuous use of the road. In 1989, the North Carolina Court of Appeals upheld the ruling.

Harrell Paden was one of those who testified in the court battle. He said his fishing days on the west end of Holden Beach were numbered, but that he was fighting on behalf of others, like "the people who live in Whiteville, fifty miles inland. They don't own a house on the beach, but they'd like to come down here."

Paden was a fisherman's fisherman. He owned twenty-four rods, bought hooks by the hundreds, and made his own sinkers. One day, he talked his brother-in-law into joining him at the west end of Holden Beach. They hooked their lines after dark—the best time to catch puppy drum—and waded into the surf. "I was standing about seventy-five yards away and I heard him yelling," Paden said. "I thought somebody was killing him, the way he was going on. His drag was a-singing. He brought in a ten-pounder and from then on there wasn't but one place to fish, and that was the west end of Holden Beach."

RAKING FOR CLAMS IS THE ONLY LIFE FUZZY SPIVEY KNOWS.

PHOTOGRAPH BY DEIDRA LAIRD

The Only Life They Know

FUZZY SPIVEY has raked clams in the Cape Fear and Elizabeth rivers for more than twenty-five years, his bronzed muscles and fierce independence testimony to the solitary life of a clammer. He is one of fifteen hundred men and women who harvest clams by hand along the coast of Brunswick County.

It's been a way of life for generations, and Spivey is fighting to keep it that way. He and other hand-clammers worry that mechanical clam harvesters will strip the river bottoms and force them from their jobs. They say that dredges pose a worse threat than the toxic red-tide algae that shut down 230 miles of coast in the Carolinas to shellfish harvesting in 1988.

"I'll put it to you blunt and plain," Spivey says. "If they put those dredges in here, you might as well forget about clamming in Brunswick County. It'll ruin us."

Hand-clammers oppose mechanical harvesting for two reasons. They say mechanical harvesting wipes river bottoms clean, which prevents clams from replenishing and destroys other marine life. They say it also floods the market, reducing the price per clam.

Mechanical harvesters are just as adamant. In 1988, they petitioned for the firing of Bill Hogarth, the director of the North Carolina Division of Marine Fisheries, because he wouldn't open new areas to them. Some hand-clammers threatened to sink dredges if he did.

"It's a very sore subject in this county," says Sandy Tyner, a Southport seafood dealer who buys from Brunswick County clammers. "Being clammers, they know that if dredge boats are allowed in, it will cut their business off. In thirty days, there won't be enough clams to make a living."

Hand-clammers gather five hundred to a thousand clams in a good day. Mechanical harvesters get that many in twenty to forty minutes.

Hand-clammers use several time-consuming methods. Some dig sand bars with short rakes that look like potato rakes. Others drag shallow-water bottoms with bull rakes equipped with

RANDY CURTIS HARVESTS CLAMS WITH A HAND RAKE.

PHOTOGRAPH BY DEIDRA LAIRD

teeth to dislodge clams and baskets to catch them. Still others work from their boats with sixteen-foot tongs.

In shallow water, mechanical harvesters spin their boat propellers along the bottom, driving clams into a net behind the boat—a process called kicking. In deeper water, they use hydraulic dredges to suck clams onto a conveyor belt.

Rusty Mills works up the coast at Crab Point, near Morehead City. Mills has kicked for clams for ten years. He'd like more areas opened to mechanical harvesters. His main concern is that existing areas not be closed. Mills concedes that mechanical harvesting could destroy the Brunswick County hand-clammer.

"If we were allowed to go anywhere that a raker could go, it probably would close them right out," he says. "Our means is more efficient than raking. We could go make a day's work where they couldn't find one. But most people kicking . . . ain't trying to close them out. We're just hoping . . . we end up with what we've been having. I wouldn't look for any further bottom to be open. Whenever you try to get some new bottom, you got this side that don't want it, and most time we're outnumbered."

Clamming is a $7-million industry in North Carolina. There are 150 mechanical harvesters, compared with 14,000 hand-clammers. But in the four months mechanical harvesting is allowed—December through March—mechanical clammers gather about a third of the year's harvest—414,044 pounds, compared with the 942,272 pounds brought in by hand.

Fuzzy Spivey says that harvesting by hand is the only living most hand-clammers know. "Mechanical harvesting is stupid," he says. "And there weren't nary one to tell us how we'd gain from it in Brunswick County."

IF MECHANICAL HARVESTING IS ALLOWED, HAND-CLAMMERS SAY THEY'LL NO LONGER BE ABLE TO MAKE A LIVING.

PHOTOGRAPH BY DEIDRA LAIRD

TWO VIEWS OF OLD BALDY,
WITH NATURALIST BILL BROOKS
IN THE FOREGROUND BELOW

PHOTOGRAPHS BY BOB LEVERONE

BALD HEAD ISLAND

. . .

TUG OF WAR

FOR TWENTY YEARS, environmentalists and developers fought bitterly over Bald Head Island—its marsh grass and coastal creeks, lonely stretches of sand, nesting loggerhead turtles, great white egrets. One side wanted a state park, the other a resort.

By the time Kent Mitchell first visited in 1983, environmentalists had lost the battle. Mitchell thought the land should be preserved, but he decided that if the island was going to be developed, he was the one to do it.

"Maybe Bald Head should have been a state park," said Mitchell, an architect from Texas and president of Bald Head Island, Ltd. "But the state legislature voted against it several times. . . . I'm doing the best I can. . . . I'm trying to put development on the areas least sensitive to development, and those most sensitive—most vulnerable—I'm trying to leave alone."

Even people who fought to keep Bald Head unspoiled concede that development on the island has been unlike the destructive buildup on much of the coast of the Carolinas. The once-desolate island south of Wilmington now boasts an eighteen-hole golf course, croquet greens, and hundreds of houses and condos. "I would have preferred to see it left alone, but given development, I think they've done pretty good," said James Parnell, a University of North Carolina-Wilmington biologist. "They don't let automobiles on the island and they have not put in a lot of high-density stuff."

But a colleague of Parnell's at UNC-Wilmington, biologist David Webster, warns that any development ruptures the natural rhythm of the coast. "The developers who are developing the island have a guilty conscience," Webster said. "They try to make decisions that are environmentally correct, and I think they've accomplished their goal in large part. But you've still got a conflict with nature. . . . They're impacting in a negative way on one of the most important ecosystems in the world. It's prime real estate and these guys could not ignore the lure of the mighty dollar."

For nearly two hundred years, Bald Head

stubbornly resisted development. The island sits at the mouth of the Cape Fear River, twenty-five miles south of Wilmington. It is home to more salt marshes and nesting loggerhead turtles than any other place in North Carolina. Bald Head owes its name to its highest dune, worn bare during colonial days by sailors who climbed to the top to spot incoming ships. Bald Head's beaches stretch for fourteen miles. In the heart of the island, where pirates Blackbeard and Stede Bonnet are rumored to have buried treasure in the 1700s, a maritime forest features a wild mix of oaks, loblolly pines, wax myrtle, and yaupon.

Over the years, many people tried to tame the island. Most failed. One of the first was North Carolina Governor Benjamin Smith, who dreamed of settling Bald Head in the early 1800s but ended up in debtor's prison. One developer after another followed, including Frank O. Sherrill of Charlotte, who boasted in the late 1960s that he would build a bridge to the island from Southport, several miles across the Cape Fear River.

Governor Bob Scott urged the state to buy the island for a wildlife refuge in 1971. The North Carolina General Assembly refused. One developer gave nine thousand of the island's twelve thousand acres to the state for a nature preserve, but most of that land was salt marsh that the state already claimed. Lots finally began to sell in the early 1970s, and construction began.

In 1976, Tar Heel native and Texas millionaire Walter Davis bought Bald Head Island. Seven years later, he sold it to his friend George Mitchell, a Texas energy executive, and Mitchell's sons, Kent and Mark. The Mitchells paid $15 million to $20 million for an inn, a restaurant, a golf course, and about sixty houses and villas.

"What took my breath away was its beauty, a sense of natural wildness," Kent Mitchell said of the island. "We've had to make compromises. That's the nature of development. But we've tried to strike a balance between the natural and man-made environment, between making money and maintaining the natural environment."

Instead of streets, narrow asphalt paths wind their way among dunes and oaks. Instead of cars, electric-powered golf carts provide most of the transportation. Property owners created a nonprofit conservancy that pays for a full-time naturalist. Part of the job is to judge the effect of development on the island. When road workers began to lay asphalt near a large oak, naturalist Bill Brooks persuaded them to shift the path and spare the tree.

There are no streetlights on Bald Head Island, and no hotels. The Mitchells brag that the island's only high-rise is Old Baldy—at 179 years, the oldest lighthouse in a state famous for its lighthouses. There's also no bridge to the island. A ferry, at $28 per round-trip in 1991, discourages tourists from going to the island, where lots sell for up to $250,000. Only a few dozen people live on Bald Head year-round. As many as a thousand go there on summer holidays.

"What makes Bald Head unique also makes it a little difficult," Mitchell said. "It's hard to

Bill Brooks Uses His Telescope To Check On Waterfowl.
PHOTOGRAPH BY BOB LEVERONE

get to. It doesn't have a boardwalk full of video games. It doesn't have thirty different restaurants. A person who's going to buy there probably bought at Pawleys Island thirty years ago. It's a special place. We're trying to keep it that way."

But for all the planning, there have been pitfalls. On December 31, 1988, islanders hosted a farewell party for Bald Head Island's inn. Built in 1972, the inn was moved three hundred feet inland in 1980 because of erosion along the shoreline. That wasn't far enough. The inn was literally in the water—with waves washing up and under the building—by the time of the farewell party.

William Moss, a Greensboro executive, owned two lots on the island worth $148,000. A northeaster hit on January 1, 1987, taking with it most of his property. Moss described his investment as "a fishing hole."

Such losses should serve as a warning to developers, said David Webster. "Despite the fact that your intentions are good, you can only mess with Mother Nature so much."

Robert Harrell Found Happiness As The Fort Fisher Hermit.

Charlotte Observer file photo

. . .

The Hermit Goes Home

THE FORT FISHER HERMIT is back home.

Seventeen years after he was buried in Shelby, in the foothills west of Charlotte, Robert Harrell was returned to his beloved seashore. On June 4, 1989, friends and relatives reburied his body in a shell-covered grave in an old Carolina Beach cemetery, not far from the abandoned World War II bunker where Harrell camped from 1955 to 1972, living off the land, the sea, and spare change from curious beachgoers. The inscription on his tombstone reads, "He Made People Think."

"He never wanted to leave Fort Fisher," said Gaile Welker of Greensboro, who researched the Hermit's life for a book. "He told a lot of people not even to bury him—just leave him down there and let the crabs eat him. He found a type of acceptance at the beach that he didn't find in Shelby. That's where he belongs."

For longtime coastal residents like Harry Warren, the Hermit was as much a part of the old Confederate fort as sand dunes and fiddler crabs. "I met him when I was a little boy," Warren said. "I was pretty intimidated by him, an old man living way off down there. He had a big old iron frying pan out front with a few coins in it. My father gave him a couple of dollars. In return, the Hermit always gave you something. He gave me a piece of shrapnel. He knew a little boy would like that."

When he moved to Fort Fisher, a desolate spit of land fifteen miles south of Wilmington, he was sixty-two years old. His wife and three children had left him in the late 1930s, tired of living a gypsy life. Harrell had a temper and a history of mental illness. He had earned a living—if you could call it that—hammering out key chains and jewelry on sidewalks or working as a newspaper typesetter.

At the beach, he found a peace that had eluded him. He took squatter's rights to an old gun emplacement on the edge of a salt marsh. He drank rainwater coffee and dined on oysters, shrimp, and fish gathered from nearby Buzzard's Creek. Most of the year, he went bare-chested and barefoot, dressed in a bathing suit and a ragged straw hat. His hair and beard grew wild. Over the years, he stooped hunchback, shorter than his natural five-foot-three. The sun aged his skin so tough that he didn't feel the deer flies landing on his back.

At first, islanders and vacationers didn't know what to make of him. "Then they saw that he was harmless," Harry Warren said. "When his reputation grew, people came down here by the hundreds to see him. . . . The folks of Pleasure Island adopted him as a leading tourist attraction."

The Hermit boasted as many as seventeen thousand visitors every year. Those who met him were struck by his intelligence. He talked easily with them about politics, religion, the coast.

"I came here to write a book, my think-book on humanity," he said in 1971. "I'm here for the same reason Dr. Schweitzer went to Africa. There's no difference in our goals. His was medicine and missionary work. Mine is psychology."

Mary Sessoms of Carolina Beach remembers one piece of psychology. "He would put a fifty-cent piece, a quarter, a nickel, a dime, and a penny in that iron frying pan in front of that old shack thing he lived in. He told me he did that every morning. He always said, 'I'm using psychology on people.' He said anybody who came there knew that a penny was welcome, a nickel was welcome, a dime, a quarter, or a half dollar."

The Hermit buried the money in old jars in the sand. Some people believe there's still some buried treasure.

Harrell died on June 3, 1972, at age seventy-nine. His death was listed as due to natural causes. But his son, Ed Harrill—the Hermit changed the spelling to Harrell—thinks robbers killed him. When his body was found, his raincoat was bunched around his neck, his legs were bloody. There were shoe prints in the sand and marks that suggested someone had dragged the body. At Ed Harrill's request, the body was exhumed in 1984. An autopsy showed no evidence of foul play.

"That was his home and his glory," said Harrill, a retired machinist and coal miner from Chattanooga, Tennessee. "I did not want my father living like that. But he loved it. I first went down to see him in 1956. I told him I didn't want him living there.

"He looked at me and he said, 'What are you crying for?'

"I said, 'Because my father's living here like that.'

"He shooked that little finger in my face—it ain't the first time he did. He said, 'You're up there in Cleveland, Ohio, killing yourself for millionaires, trying to raise your nine kids, having a hard time of it. I'm living here free. I got it made, like I'm in heaven or something.'

"I never cried no more."

FRIENDS REBURIED THE HERMIT AT THE BEACH IN JUNE 1989.

PHOTOGRAPH BY JIM GUND

THE ELWELL FERRY TAKES OVER WHERE THE HIGHWAY ENDS.

PHOTOGRAPH BY JIM GUND

CROSSING THE RIVER

MOST PEOPLE end up at the Elwell Ferry on purpose. Those who don't are in for a surprise.

The look on motorists' faces when the blacktop on State Road 1730 runs into the Cape Fear River always sets Bob Mitchell to laughing. "They don't know what to make of it," he says. "It's a lot different from modern-day transportation."

Instead of a bridge, a ferry takes motorists across an isolated stretch of the Cape Fear thirty miles west of Wilmington, near the community of Kelly. The operator of the ferry lowers a ramp to shore so motorists can drive aboard. There's enough room for two cars bumper to bumper or one car with a trailer. With the flick of a lever, the operator starts the diesel engine that powers the propeller. Two steel arms connect the ferry to a cable, which guides the ferry a hundred yards across the Cape Fear.

The free trip takes ninety seconds. It saves thirty minutes.

Without the Elwell Ferry, Vance Callihan would have to drive through Elizabethtown—miles out of his way—to get from his home in Bladenboro to his favorite fishing spot on the Black River. "I've rode it ever since I was a teeny little boy," Callihan says. "It saves a lot of miles."

The Elwell Ferry is a holdover from the days when crossing rivers was a major obstacle to travelers. There are only two other inland ferries left in North Carolina, both in the northeastern part of the state—the Parker Ferry, which crosses the Meherrin River near Murfreesboro, and the Sans Souci Ferry, which crosses the Cashie River southeast of Windsor. The Elwell Ferry is the only one that crosses a navigable river.

Residents have come to love the Elwell Ferry like an old friend. It's been there—between N.C. 53 at Kelly and N.C. 87 at Carvers—since 1905. In the early days, brothers John and

Walter Russ used poles to propel the ferry and its cargo of mule-pulled wagons. The first motorized car crossed on the ferry in 1916. Back then, automobile brakes didn't work as well as they do now. At least one car drove onto the ferry and kept going right into the muddy Cape Fear. The driver got out safely, but the car had to be brought up piece by piece.

The ferry was motorized in 1939. Three years later, its gasoline engine misfired, destroying the ferry and killing Walter Russ.

Today, operators must train for 180 days and pass Coast Guard certification requirements to run the ferry. Christmas is the only scheduled day off, but the ferry doesn't run in fog or if the river swells its banks.

"You meet interesting people," says Mitchell, a farmer who operated the ferry for more than ten years. He recalls a man who came to the ferry boasting that his dog would do anything he did. The man set out to prove it. He jumped onto a rope swing located nearby and swung out over the Cape Fear. When he got down, the dog jumped up. "There was that dog turning around and around," Mitchell says.

On an average day, seventy-five cars cross the river on the ferry. Some passengers come out of curiosity. Most come out of necessity. They are farmers and fishermen who live in the area.

The ferry operators record each passage on a pad in their four-by-six-foot control room. June is the busiest month, with about a hundred cars a day. That leaves little time for listening to the radio or enjoying the countryside.

Just when things seem to be getting dull, a limousine will cross. Or a deer will swim by.

At the end of the day, the operators dock the ferry on the northeast side of the Cape Fear. They drop the cable twenty-four feet to the bottom of the river so barges can pass safely in the night. That's when L. W. Porter uses the ferry. He fishes off it. Sometimes other fishermen join him, angling earthworms into the deep water.

"I caught a forty-one-pound catfish off the ferry," says Porter, a retired newspaper deliverer. "One time I got a four-pound catfish that jerked the line broke. The catfish swam toward the ferry, and I reached over and got the line and pulled him in like that. It's a good place to fish."

Every now and then, Porter also uses the Elwell Ferry to cross the river.

MASONBORO BOATYARD

. . .

THE LIVE-ABOARDS

MARY JO BENNETT calls herself a live-aboard. Hundreds like her live up and down the coast, people who traded manicured lawns for mud flats at low tide and playful porpoises at high tide.

She and her husband, Dick, stopped at Masonboro Boatyard in 1972 when the alternator shaft broke on their thirty-two-foot sloop. Dick Bennett died at the boatyard in 1991. Mary Jo hasn't left yet.

Masonboro Boatyard sits at the mouth of Whiskey Creek, just a short tack from the Intracoastal Waterway. Wind-swept oaks line the shore, and sailboat masts line the wooden docks. Mary Jo Bennett's 250-square-foot houseboat, the *Last Gasp*, is moored alongside several others that look from a distance like tidy homes along a narrow neighborhood street. One has a shingled roof and miniblinds. Another, a stained-glass window. A third, a front-door wreath.

Mary Jo Bennett plants tomatoes, basil, parsley, oregano, chives, lettuce, and New Zealand spinach on the deck of her houseboat. Inside, there's room for a television, microwave, coffee maker, two woks, a deep-fat fryer, food processor, air conditioner, ceiling fan—even a silver ceremonial sword from Indonesia. The houseboat offers all the comforts of home—albeit a small home. The boat is the size of the Bennetts' living room back in the mid-1950s, when Dick taught economics at Whitman College in Walla Walla, Washington, and Mary Jo taught third grade.

"Live-aboards are weird, certainly unconventional. But I prefer weird," Dick Bennett said in June 1988 from his customary perch on the sofa bed. He hadn't been off the dock in two months. "I'm a curmudgeon. I quit work when I was forty-six. . . . We quit as soon as we got enough money that, I figured if we were careful on spending, we could live on the dividends. I built this here [the houseboat] in 1982. We weren't cruising anymore. . . . We've had three cruising sailboats and have gone all around the world, from the Pacific Northwest within fifty

Dick And Mary Jo Bennett And The *Last Gasp*

PHOTOGRAPHS BY DEIDRA LAIRD

miles of Alaska to the Java Seas. We were cruised out. There's not a lot of room on a thirty-two-foot sailboat."

Ross Mann can attest to that. He lived on a thirty-seven-foot sloop at Masonboro Boatyard for more than three years.

Like the Bennetts, Mann plotted and pinched his way through a nine-to-five job. As a school principal at Sandhills Youth Center and Cameron Morrison Youth Center, near Southern Pines, he always dreamed of life on the sea. He bought his sailboat in the mid-1980s and visited Masonboro Boatyard on weekends for three years before moving to the boatyard.

Mann "lived on the hard," meaning that he kept his boat on land instead of in the water. He repaired the blistered fiberglass hull himself. He redid the interior woodwork—an elegant blend of dark teak and light ash—and added a tape player and four speakers. He showered at facilities in the boatyard, swam out Whiskey Creek to the Intracoastal Waterway for evening dips, and watched "L.A. Law" and "Star Trek: The Next Generation" on an eight-inch television by his bed in the bow.

"I picked this boatyard because it has a reputation as a do-it-yourself boatyard," he said. "You don't have to wear whites."

DICK AND MARY JO BENNETT WOULDN'T LIVE ANYWHERE ELSE.

PHOTOGRAPH BY DEIDRA LAIRD

For about ten years, Dick and Mary Jo Bennett borrowed a car when they needed to go into nearby Wilmington. Mary Jo turned sixty-two in 1984, and they spent her Social Security check on a used Toyota Celica. In 1985, they finally hooked up a telephone.

There are few problems in Masonboro Boatyard. The worst Mary Jo Bennett has faced is keeping her footing on the tiny deck of her houseboat. Once, she leaned too far back and went overboard. Her cat did it, too.

"It bangs some in storms," Dick Bennett said of his home. "In hurricanes, I just leave. Most of the time, I'm sedentary. I watch the damn birds. I like the otters better. They play. We just saw a green heron, and there's a mallard duck on that bank over there.

"What more could I ask for?"

ROSS MANN TRADED A JOB AS SCHOOL PRINCIPAL FOR LIFE AT SEA.
PHOTOGRAPH BY DEIDRA LAIRD

Like a Row Of Homes On A Street, Houseboats Float Side By Side At Masonboro Boatyard.

PHOTOGRAPH BY DEIDRA LAIRD

THE DRAM TREE

FOR THREE HUNDRED YEARS, sailors took a swig of rum whenever they passed a certain cypress growing in the black waters of the Cape Fear River. They called it "the Old Dram Tree."

As they floated past on their way out to sea from Wilmington, the sailors would roll out a barrel and take a good-luck drink—a dram—before setting sail in the Atlantic. On their way back, they toasted their safe return.

The Old Dram Tree was as laden with legends as it was with Spanish moss. One legend has it that when the pirate Stede Bonnet was pillaging the Atlantic coast in 1718, he sailed up the Cape Fear one balmy summer afternoon in search of supplies. He dropped anchor at the plantation of a pirate turned planter. While the planter tended to business a few miles away, Bonnet and his crew raided the liquor cabinet. Drunk, they kidnapped the planter's wife and sailed off before her husband returned.

Once under way, the pirates untied the woman and busied themselves with their ship. When they passed the Old Dram Tree in the early evening, the woman quietly slipped from the stern into the water and swam to the tree. As the legend goes, she hid in its lower branches and was still there, safe, when her husband came looking for her the next morning.

More than fifty years later, on a cold and rainy day during the Revolutionary War, a British officer is said to have jumped ship and hid in the same spot. After a fruitless search, his ship sailed away, and the officer swam ashore to join forces with the colonists.

The Old Dram Tree was a fixture in local lore until forty years ago, when some port workers unwittingly cut it down.

"During three centuries and more of existence, this stately cypress towered as a guardian, welcoming the incoming mariner and bidding him farewell," Louis T. Moore wrote in his history of Wilmington. "The hoary old veteran of the forests was well protected from storm and wind and from time immemorial had withstood the ravages of the elements. It was tragic indeed when it was pulled down by careless workmen, who did not care."

In 1989, in celebration of the 250th anniversary of the city of Wilmington, local residents planted a new Dram Tree. "We're just doing the tree bit," said resident Jimmie Davis, who suggested planting the tree and continuing the tradition. "The navy got away from the free grog a long time ago."

The Old Dram Tree stood sentinel fifty feet offshore, about two miles south of Wilmington. The new version was planted at the foot of Castle Street, near the Cape Fear Memorial Bridge. It is both a birthday salute and a symbol of renewed interest in historic Wilmington. The city began to slump in the 1950s, after Atlantic Seaboard Railroad pulled out of town. But in recent years, there's been revitalization.

"When it was cut down, it was kind of the end of Wilmington. The city was going down," said artist Claude Howell. "By then, the Dram Tree was just a half-dead cypress with a few branches and a lot of moss hanging down. It was quite noticeable because it was out in the river. In those days, people were tearing down everything in Wilmington.

"Now, they've put the Dram Tree back up, and Wilmington is booming again."

JIMMIE DAVIS THOUGHT WILMINGTON NEEDED A NEW DRAM TREE.

PHOTOGRAPH BY BOB LEVERONE

Bill Hurst Feels At Home And Happy On Masonboro Island:
"When I Get Here, I Don't Worry About Anything Anymore."

PHOTOGRAPH BY DEIDRA LAIRD

The Fight For Masonboro Island

ELIJAH HEWLETT knew a good thing when he saw it.

Hewlett bought tiny Masonboro Island before the Civil War so he could fish for mullet. Back then, there were islands like Masonboro up and down the coast, deserted stretches of sand undisturbed except for an occasional fisherman.

Now, Masonboro Island, ten miles from downtown Wilmington, is the only undeveloped barrier island on the southern part of the North Carolina coast. Multistory buildings rise out of the sands of Wrightsville Beach to the north and Carolina Beach to the south. The highest point on Masonboro is a dune. The only "building" on the island is that of the sand crab, the loudest noise the cry of the tricolored heron.

More than a thousand residents in the Wilmington area have fought to keep it that way.

"My concern was that for many, many years the barrier islands have been slipping away," said Edith Friedberg, who organized the nonprofit Society for Masonboro Island in August 1983 after reading an advertisement for oceanfront lots on the island. "I think nature is less replaceable than people. When something's gone from our natural world, there's no reproducing. I'm always scared about the land. I'm not a scientist. I'm not schooled in things environmental. But in my life, there's nothing more important than Masonboro Island."

Friedberg is a relative newcomer. She's been visiting Masonboro Island for twenty years. Others have gone there for generations. Though privately owned until the late 1980s, the island was open to anyone who could get there by boat.

Masonboro Island is nine miles long and scarcely wider than a football field in places. For the most part, it is too low-lying for development. Elijah Hewlett's descendants inherited a large portion of the northern end, suitable for building houses. There, wind-swept dunes overlook a crescent-shaped beach to the east and the

Intracoastal Waterway to the west.

Wilmington lawyer Addison Hewlett spoke for most of the hundred-plus descendants of Elijah Hewlett when he said in 1988, "We'd like to see the state buy it. My grandfather was a farmer down on Masonboro Sound and he also was interested in fishing. He had a seine crew that fished for mullet over on the ocean side of the beach, and that was what he wanted it for back then. . . . Folks have gone over there for years. It's been open to anyone. But if it was developed, you know how that would be."

The state of North Carolina began buying lots on the island in 1985. By 1991, it owned three-fourths of the land, now part of the National Estuarine Reserve Research System and protected like Zeke's Island, near Fort Fisher; Currituck Banks, on the northern Outer Banks; and Permuda Island, near Topsail Beach.

Bill Hurst, another of Elijah Hewlett's descendants, began camping on the island as soon as he was old enough to paddle a johnboat. He wanted all of Masonboro Island to go to the state, and he thought Elijah Hewlett would have wanted that, too.

"Have you ever seen anything prettier in your life?" Hurst asked one morning shortly after dawn as he headed toward the island. A three-pound Spanish mackerel jumped out of the Intracoastal Waterway, its forked tail glistening in the morning sun. A mullet left a wake near the green marsh grass. Pelicans flew over the surf nearby.

"There are areas like Figure Eight Island where you and I cannot go," Hurst said. "In time to come, if something isn't done, there will be a whole lot of poor and middle-class people who have no hope of experiencing it.

"When I was a kid in grammar school, all the boys couldn't wait until school was out. We would get a big old church tent, set it up, and everybody would bring cans of pork and beans and some colas, and for a week or two we'd have a delightful time. . . .

"A lot of people come over here. There are no restrictions. You can build a tent, go barefoot, chew tobacco. This used to be a good place for lovers, too. . . .

"When I get here, I don't worry about anything anymore."

EDITH FRIEDBERG

PHOTOGRAPH BY DEIDRA LAIRD

As a Child, Bill Hurst Spent Weekends
Camping On Masonboro Island.

PHOTOGRAPH BY DEIDRA LAIRD

The Menhaden Chanteymen Ride On One Of The Fishing Boats.

Photograph by Gary O'Brien

THE MENHADEN CHANTEYMEN

ELEVEN MEN step single-file into the dim parish hall, all slowed by age, a few stooped from years of hauling fishing nets.

They doff their caps and sit at two wooden tables. A few pleasantries. Some business. Then, without a cue, John Jones begins their Friday-evening ritual, his strong tenor filling the small room: "Oh, I'm gonna roll here."

A chorus of ten harmonizes the refrain, synchronizing voices the way they used to synchronize pulling in the nets: "*Roll here a few days longer.*"

"I'm going to roll here," Jones continues.

"*Roll here a few days longer.*"

"Then I'm going home, boys."

"*Lord, Lord, I'm going home.*"

In their minds, they are back on the water, ocean waves rocking their wooden purse boats as they struggle to raise their net against the weight of a half-million pounds of menhaden, fingers clawing at the mesh, shoulder and back muscles straining, blood vessels breaking in overworked arms.

They sing a verse. They raise the net a foot.

A verse. A foot.

Verse by foot, they lift their catch of small, silvery fish to the surface.

They call themselves the Menhaden Chanteymen of Beaufort, North Carolina, as fabled for the chanteys they sang as the fish they caught. Thirty years after the hydraulic lift took the place of their muscles, the last of the Menhaden Chanteymen reunited. With encouragement from a folklorist, they are preserving a dying and beloved part of black history in the coastal Carolinas.

"Mornings, you know, we used to fish, it sounded like a church on the water," says Harvey Frazier. "It sounded so good, people passing by with boats would stop and listen to the singers pulling in nets."

From the 1880s to the 1950s, menhaden

boats lined the Beaufort waterfront, bow to stern along Taylors Creek and on up to Morehead City. More men caught and processed menhaden than any other fish in the Carteret County area, and Beaufort processed more menhaden than any other place in the United States. As the saying goes, those were the days when "boats were made of wood and men of steel."

The men who made their living off menhaden called the fish by more familiar names—shad, porgies, fatback. Known as the soybean of the sea, menhaden are stinky, oily fish about the size of sea herrings, no heavier than a pound and no longer than fifteen inches. Bluefish feed on them. Some people fry them. But menhaden are valued chiefly for their oil, used to make cosmetics, paints, linoleum, livestock and poultry feed, antibiotics, and five hundred other products. What's left is fertilizer.

Fishing for menhaden was a calling. William Henry began as a teenager. His father was John "Coomy" Henry, a legendary chanteyman whose name evokes memories of stirring chanteys sung with a gospel flavor. "When some of the men would be out, and get drunk that night, I'd fill right in," William Henry says. "I was sixteen years old, a boy. I pull so hard the veins in my arms jumped up."

Henry, still trim and muscular, stands six-foot-one and weighs 150 pounds. Only his graying temples hint at his age or the hard work.

Before dawn every morning, as most of Beaufort slept, Henry would take his place among two dozen other fishermen as the big menhaden boat put to sea. Sixty-five feet up in a crow's nest, a spotter scanned for menhaden schools, which could run miles long and wide. He looked for telltale signs: the splash of tails as the menhaden cut the water, or a dark red cloud on the surface.

"Fish! In your boats!" the spotter would call.

Twenty men divided into two small crafts known as purse boats, one piloted by the captain, the other by the mate. As the men heaved the sixteen-foot oars, the boats plowed the water side by side, joined by a huge net tied to both boats. "Let them go apart!" the captain shouted as the boats neared the fish. The two crews rowed in opposite directions, the net spreading between them. The boats cut a semi-circular path until they met again, the net behind them in a circle.

At the captain's cue, the strongest man pushed overboard a seven-hundred-pound weight attached to the net. Like a drawstring, it gathered the bottom into a purse and trapped the menhaden. The mother ship would join the two purse boats, forming a triangle with the net of menhaden in the middle.

The work began.

At first, it was easy to hoist the net. It came up yard by yard. But as the fish bunched together in the bottom, the men felt the full weight of their catch. They would lean back, feet against the ribs of the boats, and the lead singer would begin, "Won't you help me to raise 'em, boys?"

"*Oh, oh, honey.*"

"Will you help me to raise 'em, boys?"

"*Oh, oh, honey.*"

"Will you help me to raise 'em? I'm going

Two Boats Of Menhaden Fishermen Ready To Lift Their Catch.

COURTESY OF N. C. MARITIME MUSEUM

They Once Sang To Synchronize Pulling In Their Fishing Nets. Now They Sing For Enjoyment, Followed By A Prayer.

Photograph by Gary O'Brien

to . . ."

"*See her when the sun goes down.*"

The verse over, they pulled the net a foot.

"Oh, the weight's on the mate boat."

"*Oh, oh, honey . . .*"

The chanteys synchronized the pulling of the net and gave the fishermen a breather. Some were gospel tunes, some were spirituals, and some were work songs. As with farm workers picking cotton or railroad workers laying track, the fishermen drew the strength and will to complete their work from the songs they sang. A good chanteyman was said to be as valuable as a good fisherman.

The fishermen made two cents for every thousand pounds caught. Some weeks that meant as much as four hundred dollars, other weeks nothing. John Jones raised five children—three girls and two boys—and sent them to college on his fisherman's wages. "I told them, 'It's rough out there on the water,'" he says. "It wasn't no life for them. I've seen so many men lose their lives, seen so many boats get lost."

Jones retired in 1982 after thirty-five years on the water. He quit fishing, and he quit singing chanteys.

Then, one day in 1988, folklorist Michael Luster asked him to sing. Luster, from Irving, Texas, had come to Beaufort with his wife and partner, Debbie, on a six-month project for the North Carolina Arts Council. They heard about the old-time chanteys, and when they listened to Jones, they were captivated. They recruited other retired fishermen—most of them in their sixties and seventies—to sing at a local festival.

"The chanteys were gone, the industry itself was just about gone," Michael Luster says. "This was a working waterfront town for a long time. Now, it's a stop on the Intracoastal Waterway. I thought it important they remember the working waterfront."

The fishermen enjoyed their performance as much as the community did. They wanted to keep on singing.

In 1990, the Menhaden Chanteymen recorded a cassette called "Won't You Help Me to Raise Them," and sang at Carnegie Hall in New York City.

At home in Beaufort, they are helping restore local pride in the industry that gave the town its first name—Fishtown—and supported families like theirs for nearly three hundred years. Every Friday night, they come to the parish hall at St. Stephens Congregational Church. They sit on wooden folding chairs or green plastic chairs under the dim glow of a bare bulb. Passersby stop to listen just as they did in the old days, when yachts would still their engines to hear the haunting serenade.

For forty minutes, the Menhaden Chanteymen lose themselves in their songs. Some keep the beat with their hands, others pantomime pulling the net. In their minds, they are back at work, back at sea.

"How can I go home?"

"*Got no ready-made money.*"

"How can I go home?"

"*Got no ready-made money.*"

"To settle down, boys."

"*Lord, Lord, to settle down . . .*"

Grayden Paul, Beaufort's Most Beloved Historian

Photograph by Bob Leverone

In His Mind's Eye

GRAYDEN PAUL pointed to his left and asked, "You see that great big oak tree over there? The one with the great big limb about twenty-five feet long with a bend just like my elbow? That's where they hung him." Then he spun a tale about the hanging tree.

But Beaufort's most beloved historian couldn't see what he was talking about. For the last half of the thirty years he conducted tours of town, Grayden Paul was blind.

Riding through Beaufort on a double-decker tour bus, his cane and a thermos of water close by, Paul urged tourists to "look there" and "see here." When they passed a historical marker, he would repeat the notation nearly word for word.

In his memory, Paul saw as clearly as if it were 1960, the year he organized the Beaufort Historical Association and gave his first tour. Even back then, Grayden Paul knew more about Beaufort than the history books.

He was born in 1899. His roots in Carteret County date back to the Civil War, when Federal forces captured his grandfather, Raymond Luther Paul, and put him in prison. Grandfather Paul caught the eye of a nurse's aide, who slipped him sandwiches through the bars of his cell. Once free, he married her. "And that's how I got sprung here," Grayden Paul said.

His father was a jack-of-all-trades who sometimes worked as a blacksmith and other times as a tour-boat operator along Core Sound. Most days, he ran a traveling motion-picture show. Young Grayden and his five brothers and sisters tagged along as part of the act. "Momma played the autoharp," he said. "Poppa played the fiddle. My oldest sister played the piano. I came in doing the singing."

History was Grayden Paul's favorite subject from the time he was a boy. He thrived on his father's tales of pirates and haunted houses, of lost love and shipwrecks. "I was just born and raised with it," he said. "It just came naturally with me, this love of history. . . . And I don't want the history of the town of Beaufort to die.

I want to keep it alive for the other generations."

Paul made his living as a machinist. In his spare time, he wrote poems and, with the help of his wife, Mary, a history book on coastal Carteret County. "People ask me what my first love is," he said. "The first one is family. The second is church. And the third, of course, is the Beaufort Historical Association."

After 1975, when he lost his driver's license and was declared legally blind, he continued to give the tours sight unseen. In 1991, when his health failed, he finally quit. But he left behind a wealth of stories from as far back as 1709, when a group of French Huguenots founded Beaufort, the third town in North Carolina.

At "the Old Burying Ground," Paul told stories about the British officer buried standing up, saluting the king; about the little girl buried in a rum barrel after she died at sea; about Captain Otway Burns, a naval hero buried with his ship's cannon mounted on top of his tombstone.

At the 1698 Hammock House, the oldest in town, he told how sea captains brought the wood to build the house piece by piece from England. "Without contradiction, it's the first prefabricated house built in America," he would say, his face breaking into a grin.

And there was the story of how one of those same sea captains planned a party and sent for his girlfriend, along with the wives of his crewmen. The ship was still at sea when the women arrived in Beaufort by stagecoach. When the seamen didn't arrive, the women held the party anyway.

"The ship dropped anchor and the men looked over and saw every light in the house," Paul would say. "The nearer they got to the house, the louder the sound of laughter and music. The captain ran up on the front porch. That stagecoach driver was dancing with his [the captain's] girlfriend. He chased him up to the attic and murdered him. And, do you know, the bloodstains are still on the steps today and cannot be erased with all the modern detergents that money can buy. And old folks don't let their children play near[by] . . . because you can still hear that man yelling."

Though Grayden Paul no longer gives tours, he is still part of any tour of Beaufort. A drawbridge and a room in the courthouse in Beaufort are named for him. So are a recreation room at First Baptist Church and an old oak tree.

Even After He Went Blind, Grayden Paul Guided Tours Of Beaufort.

PHOTOGRAPHS BY BOB LEVERONE

GERALD DAVIS AT WORK OUTSIDE HIS SHOP

PHOTOGRAPH BY BOB LEVERONE

A Passion For Carving

GERALD DAVIS never saw a piece of scrap wood on his workshop floor. He saw two ducks.

He picked up a three-by-seven-by-twenty-four-inch stick of tupelo gum and announced, "That's two teals." When he laid the stick down, it looked like a piece of ordinary scrap wood. But when he picked up a small hatchet and began to chip, corners turned to curves. A stubby oblong became a graceful back. The wood began to soar.

Davis carved duck decoys for more than forty-five years, until he died in 1990. "I know exactly what it's going to look like from the start," he said, not boasting, just matter-of-fact. "This same bird right here, I've been making it for years and years. Almost every one turns out basically the same, though no two are alike, because they're hand-done."

In coastal Carteret County, where duck decoys seem to be about as plentiful as the real thing, Davis was a woodcarver's woodcarver. As a youngster, he whittled chain links from cedar pieces. He went on his first duck hunt at age twelve. Back then, in the 1930s, duck decoys were literally a dime a dozen. Hunters like Davis's father, Fred, made crude imitations of the canvasbacks and redheads they shot for food in Core and Bogue sounds. The men didn't pay much attention to detail. They set the wooden ducks afloat on the sounds, hoping to lure live ducks flying overhead.

Gerald Davis watched the grown men carve. Then he set about perfecting the craft. It became a passion that saw him through more than four decades of running an automobile garage in nearby Morehead City. He passed his love of carving on to Gerald Davis, Jr.—the oldest of his four children—and to three of his eleven grandchildren. He also taught about seventy people each year at a local technical college.

"All the time I was at the garage, I worked every night and weekends in the shop. I'd keep one of these in my pocket," he said, pointing to a half-finished duck. "I think anyone can take a lot of frustrations out of their mind with a piece

GERALD DAVIS TURNED WOOD SCRAPS INTO SOARING BIRDS.

PHOTOGRAPH BY BOB LEVERONE

of wood and something to cut it with."

As decoys became decorator items, Davis discovered a ready market for his work. He sold his brightly painted ducks for from forty to two hundred dollars.

Most hunters switched to plastic decoys because wooden ones are expensive. Davis remained a purist, using wooden decoys when he hunted. But as the years went by, he quit hunting as much. "The more birds I've carved over the years, the less I shoot," he said. "I think you actually feel different. I just don't like to shoot like I used to. I just think studying birds and getting closer to them is the cause of that."

Davis carved for love, not money. He owned machinery that could turn out a duck in eight minutes, but he preferred a hatchet to a band saw and grinder. "If I've got plenty of time, I'd rather cut it down," he said. "I enjoy making it this way because when I'm doing it this way, I'm doing it because I like to do it. I'm not doing it for commercial reasons."

That's why he quit hanging a nail for orders outside his shop, the Fish Town Decoy Company. "I was always making what somebody else wanted me to make," he said. "Now, I go out there in the shop, I'm going to make what I like to make."

In 1986, Davis retired full-time to his cluttered workshop, next to his house in a wine-colored barn that once held horses and dogs. The walls were covered with tools, paper patterns, and cobwebs.

As he worked, Davis tossed wood scraps onto the concrete floor in heaps. He saved the scraps to feed his potbelly stove in winter if his mutt, Termite, didn't eat them first. Chips covered the rest of the floor. "We turn out a lot of work in here," Davis said. "I don't do much cleaning."

Davis sanded some ducks to a fine, smooth finish and used handmade knives on others for such details as the barbs of the feathers. "I like to go with the flow," he said. "I enjoy any part of decoys. It makes no difference to me whether I carve a bare, basic bird or an intricate bird with all kinds of detail. Some people don't like to make but one kind. I like to make all kinds. It doesn't get boring."

In a showroom next to his workshop, Davis displayed snowy egrets, multicolored mallards, curlews, and sea gulls. After his death, people came from miles around to buy what he'd left behind.

Davis was modest. He told his students, "Anything you see here, you can make. Carving a bird like this is 95 percent know-how, 5 percent talent."

Gerald Davis had a big 5 percent.

Birds Carved By Gerald Davis Are Now Collector's Items.

PHOTOGRAPH BY BOB LEVERONE

Like Other Boatbuilders On Harkers Island, Houston And Jamie Lewis Built Their First Boat In The Backyard.

Photograph by Deidra Laird

HARKERS ISLAND

. . .

THE BOATBUILDERS

JUST ABOUT EVERYONE on Harkers Island builds boats. If they're not building boats to sell, they're building boats to sail.

Like most people on the island, brothers Houston and Jamie Lewis do it the way their father did, and his father before him. No blueprints. Lots of sweet juniper wood. Few fancy gadgets.

In maritime circles, Harkers Island means fine boats. They are recognized up and down the East Coast, distinguished by their wooden hulls and flared bows.

"Just about everybody on the island can build one," Houston Lewis says. "It's born in us. Grandparents way back built them. A lot of people don't do it full-time, but just about everybody can do it. . . . It just comes natural."

When Ebenezer Harker bought the island in 1730, he paid four hundred pounds and a twenty-foot boat for it. When Harker died, he left the island—five miles long and a mile wide—to his three sons. One of his daughters received a bushel of corn, the other a mule.

About two thousand people now live on Harkers Island, fifteen miles east of Morehead City. They've been building boats there since before anyone can remember. A handful of small boatyards operate along the island's dirt roads. At least twenty other islanders work beneath the scrub pines in their backyards.

Houston and Jamie Lewis have worked together for a quarter of a century. Jamie began building boats full-time in the family's backyard after the tenth grade.

"I started out messing with hydroplanes when I was eight or nine," he says, brushing sweat off his tanned brow. "I'd build them out of scrap lumber. My father'd make me stay and help him. I wanted to be out playing. I just sort of picked it up. I'd build one, somebody else would want one. It just got in me."

He never uses blueprints. The brothers work from a rough sketch—unless they're building a shrimp boat, that is. They've built so many shrimp boats that they don't even need a rough sketch. The brothers do their figuring in their heads. They cut planks of juniper—also known as white cedar—into one-inch strips and lay them plank by plank from the keel to the floor frames, along the curved bottom, and on up to the deck.

Their boats usually wind up bigger than the customer ordered. "The man gets more for his money," Houston Lewis says.

For fifteen years, the brothers worked out of an old shed. It was forty-six feet long, so when they built a forty-four-foot boat, they had to go out the front door, around the building, and in through the back door to get from one end of the boat to the other. Once the body of the boat was built, they hauled it outdoors to build the cabin.

In 1988, they relocated to a cavernous metal shed with a Lewis Brothers sign out front. Their boats changed along with their building. That year, for the first time, the Lewis brothers lined one of their prized wooden hulls with fiberglass. They don't like the stuff. It's messy. But it's what the customer ordered. That customer got an $80,000 fishing boat just like the $325,000 model pictured in a dusty boating magazine on a shelf in the Lewises' shed.

"We work by the hour and the customer furnishes all the material," says Houston Lewis. "The labor is the cheapest thing."

Jamie Lewis lived in a trailer until a few years back, when he sold his own boat to help pay for a house. "Most of the time, after the boat goes out, we're like we were when we started," he says. "We make grub money."

The brothers prefer building workboats for shrimpers and clammers. Nothing fancy. Nothing time-consuming. Just like the ones their father built, and his father before him.

JAMIE LEWIS SANDS A FISHING BOAT.
PHOTOGRAPH BY DEIDRA LAIRD

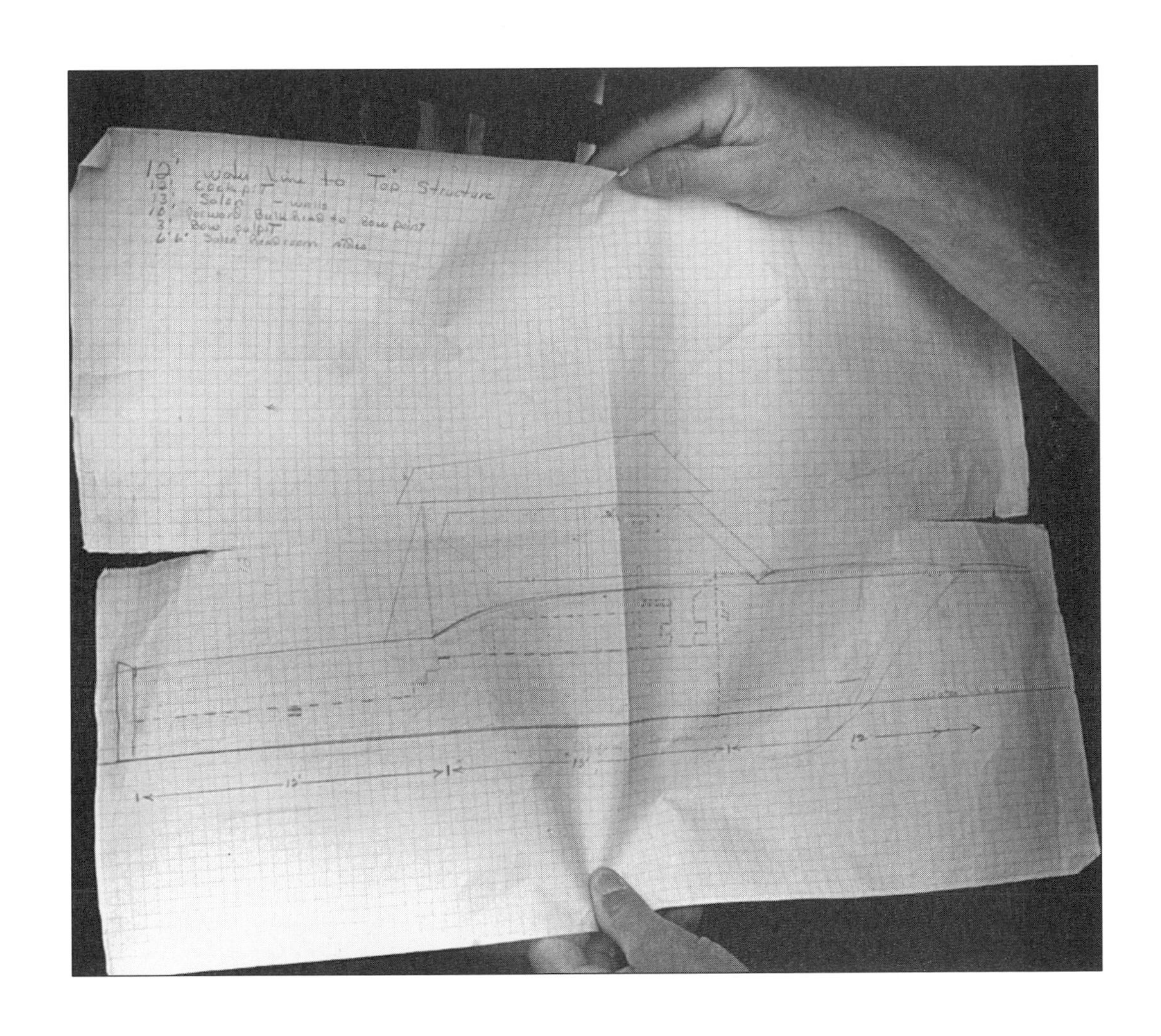

HOUSTON AND JAMIE LEWIS DEPEND ON KNOWLEDGE, NOT BLUEPRINTS.

PHOTOGRAPH BY DEIDRA LAIRD

Rany Jennette Was Born On Cape Hatteras,
The Lighthouse Keeper's Son.

Photograph by Bob Leverone

CAPE HATTERAS

KEEPER OF THE LIGHT

AS THE LIGHTHOUSE keeper's son, Rany Jennette spent a carefree childhood in the 1920s and 1930s running up and down the steps of the 208-foot-tall Cape Hatteras Lighthouse.

One day when federal inspectors were due, he decided to help his father. Jennette dipped a brush into a large vat of black liquid at the base of the tower and began painting. It wasn't paint, he quickly discovered. It was tar.

"My father came down the stairs with a leather razor strap. I knew he wasn't going to shave. He took me right over to the scene of the vandalism and said, 'Well, young man, you're going to remember this as long as you live.' I'm sixty-eight years old," Jennette said in 1989, "and I still remember it." Bits of tar remain ingrained on the granite tiers at the base of the 1873 lighthouse.

Jennette has had a lifelong love affair with America's tallest lighthouse. "There is no way of knowing how many lives this lighthouse has saved," he wrote in his memoirs, "but we do know it is truly a sentinel of the shore and a friend to seamen over all the world."

Lighthouses are no longer needed to warn sailors navigating treacherous waters. Computerized navigation took care of that long ago. Still, lighthouses hold a romantic place in our nation's history. Lonely sentinels, they stand as symbols of safe haven, of hope and perseverance. They are, as Wayne Wheeler of the United States Lighthouse Society put it, "unique altruistic structures. They stand only to serve men and women. They are kind of like a picket fence against a dark and dangerous sea."

By August 1789, sea traffic had become so great that Congress founded the United States Lighthouse Service. Today, 850 lighthouses remain. The oldest, Boston Light, was built in 1716. The newest, and probably the last, shines from Sullivan's Island near Charleston, a triangular aluminum tower built in 1960 and the only lighthouse in the United States with an elevator. All of the nation's 450 working lighthouses are automated except Boston Light.

Unaka Jennette tended the Cape Hatteras Lighthouse from 1919 to 1937. Five of his seven children, including Rany Jennette, were born in the keeper's home, which still stands in the shadow of the lighthouse. "The keeper's life wasn't as lonely as most people would think," Rany Jennette said. "The village folk always loved to visit, especially on Sunday. The life of a lighthouse keeper was a good life."

One of Rany Jennette's earliest memories is of his father tucking him under one arm and a sister under the other and carrying them up the long, dark, winding staircase to the top of the lighthouse. When Jennette was old enough to climb the lighthouse himself, he would run up the 260 wrought-iron steps as fast as he could to look through his father's binoculars at ships navigating the dangerous shoals below.

In those days, the light burned kerosene. A Fresnel lens surrounded a lantern with glass prisms that reflected the light into rays, seen as a single flash and visible for twenty miles. At sunset, Unaka Jennette lit the lantern's wick and wound the clock that controlled the speed of the rotation. The light flashed every seven seconds. Different lights along the coastline flash at different intervals, helping sailors determine their position. At Cape Lookout, seventy-five miles south, the lighthouse flashes every fifteen seconds.

At sunrise, Unaka Jennette would extinguish the wick at the Cape Hatteras Lighthouse. By day, sailors recognized the lighthouse by its black-and-white, candy-striped pattern.

The lighthouse stands by the edge of the ocean, towering over man-made sand dunes and grassy lawns where Rany Jennette played croquet as a child. His family lived there until September 1933, when a hurricane blew so fiercely that they were forced to evacuate. The tide surged around the lighthouse. Waves broke through the keeper's house, flinging aside a large oak table Unaka Jennette had propped against the door. The family never returned to live in the house.

In 1937, Unaka Jennette extinguished the beacon at the Cape Hatteras Lighthouse for the last time, because of fear that the lighthouse might slip into the ocean. A new light on a steel tower in nearby Buxton took its place.

Rany Jennette left Cape Hatteras in 1941 to join the Coast Guard. For more than forty years, he worked various jobs, from machinist to service-station manager to oceanographer. In 1983, he retired and began summer work as a seasonal park ranger at the Cape Hatteras Lighthouse, giving tours. "It's a real good feeling to be able to travel over a big part of the world and come back to within seventy-five feet of your birthplace and talk about it," he said.

Unlike others who believe the lighthouse is doomed because of erosion, Jennette believes it's there to stay. "I don't think there's a need for panic. There's not a whole lot of difference in the beach now than in the thirties. Even in the thirties, we thought it was going in the ocean. But it didn't."

SAILORS RECOGNIZE THE CAPE HATTERAS LIGHTHOUSE
BY ITS CANDY STRIPES.

PHOTOGRAPH BY DON STURKEY

SILAS DUROCHER OF FREDERICK, MARYLAND, CLIMBS ABOARD THE WEATHERED REMAINS OF THE SCHOONER *G. A. KOHLER*.

PHOTOGRAPH BY BOB LEVERONE

NORTH CAROLINA'S WATERY GRAVE

ALL THAT'S LEFT of the four-masted schooner *G. A. Kohler* is a bare-ribbed skeleton. But the ship's weathered oak beams, half-buried in the sand of the Outer Banks, evoke the days when sailing vessels defiantly rounded Cape Hatteras and plied the treacherous shoals off the North Carolina coast.

The *G. A. Kohler* was heading for Haiti with a cargo of lumber one morning in August 1933 when gale-force winds off the Outer Banks turned hurricane-force. The ship ran aground on a sand bar. Waves sprayed the deck. After a day and a half, the crew of nine men gave up hope. Miraculously, rescuers arrived. The crew fled to shore at Salvo on Hatteras Island, abandoning their ship to "the Graveyard of the Atlantic."

For ten years, the *G. A. Kohler*—one of the last cargo sailing ships built—stayed beached where the storm tossed it. During World War II, it was burned so that its iron could be salvaged. But part of the hull, buried beneath the sand, escaped the fire and resurfaced during a northeaster that battered the Outer Banks for three days in March 1989.

"Eventually, the ocean will reclaim the *Kohler* or shifting sands will bury it again," said Warren Wrenn, a ranger at Cape Hatteras National Seashore. "It's a vanishing resource, one of the things that still draws people to the Graveyard of the Atlantic."

More than a thousand ships have sunk off the North Carolina coast, most of them off the 175-mile-long Outer Banks, which extend from the Virginia state line to Cape Lookout. Sailors nicknamed the area hundreds of years before anyone thought of keeping a tally of lost ships.

Until the early 1900s, shipwrecks were a fact of life on the Outer Banks. Many locals—or "Bankers"—trace their roots to shipwrecked sailors. Many communities built their houses of wood salvaged from shipwrecks. And many an old-timer remembers the days when auctioneers sold shipwrecked cargo. Historian David Stick wrote, "It has been said of the people of Ocracoke Island, for example, that they would drop a corpse on the way to a burial if they heard the cry of 'Ship Ashore!'"

Northeasters like the one that uncovered the remains of the *G. A. Kohler* were to blame for some shipwrecks. But deadly Diamond Shoals caused most of them. The shoals are a series of sand bars shaped like a diamond and extending more than twelve miles off Cape Hatteras. "They're always shifting around," said Warren Wrenn. "You can't chart the shoals, so old navigators had a hard time of it. It's incredible how many boats were lost."

At Diamond Shoals, the northbound Gulf Stream clashes with southbound Arctic currents in a tumult of waves and sand bars. There are only two channels through the shoals, and they shift as the sand bars shift. Yet for several hundred years, ships sailing from the Caribbean to Europe went by way of Cape Hatteras. Sailors found that they could save time by riding the Gulf Stream north to the cape, then bearing east. For many, it proved a deadly shortcut. "You can be five or six miles offshore and find yourself in five feet of water," Wrenn said.

Southbound sailing ships faced other obstacles. The prevailing winds blow from the southwest. Those winds, combined with the Gulf Stream, often proved too strong for ships heading south via Cape Hatteras. Some weeks, seventy-five or more ships anchored north of the cape, waiting for the wind to shift. "A lot of captains got nervous and raised their anchors," Wrenn said. "That was a mistake, because Diamond Shoals was waiting for them."

Fragments of oak ribs and some wooden pegs in the marsh near the Cape Hatteras Lighthouse are all that's left of the three-masted cargo ship *Altoona*, which ran aground during an 1878 hurricane. "The crew thought they were going to die," Wrenn said. "But it made it through the shoals to the beach and they landed right in front of the original Coast Guard station."

The storm was so fierce, with winds of 125 miles per hour, that the Coast Guard crew didn't see the hundred-foot boat. The captain and seven men had to get down from the *Altoona* and knock at the door of the station.

No one died in the wreck of the *Altoona*, but hundreds of others lost their lives over the years. "I have a melancholy affair to relate," one captain wrote in December 1819 from Ocracoke Island. "I am the only one living of the crew and passengers of the sloop *Henry*. We left New York on Monday, 30th November. . . . Saturday morning made Cape Lookout lighthouse, . . . the gale and sea increasing so fast that we were obliged to heave to. . . . The sea began to break and board us, which knocked us on our beam ends, carried away our quarter, and swept the deck. She righted, and in about five minutes capsized again, which took off our mainsail. We were then left to the mercy of the wind and sea."

Most shipwrecks off the Outer Banks lie all but forgotten at the bottom of the ocean. A few, like the remains of the USS *Monitor*, sixteen miles off Cape Hatteras, made the history books. As ships gradually shifted from sail to steam and modern navigation and communication equipment came into use, fewer ships met their death off the North Carolina coast.

But each year, the Graveyard of the Atlantic

still claims a few.

Just before midnight on July 31, 1983, the fishing boat *Taormina* was towing the ferry *Nicolete* from Boston to Cape Canaveral, Florida. They were off Ocracoke when a thunderstorm blew in. The sea rose and the towline reeled out rapidly. The captain of the *Taormina* cut the line. When he turned back to search for the ferry, it had disappeared with a crew member aboard.

Neither man nor ferry has been seen since.

TOURISTS AT CAPE HATTERAS NEAR THE REMAINS OF THE *ALTOONA*

PHOTOGRAPH BY BOB LEVERONE

THE SALVO POST OFFICE

When mail comes to the tiny Outer Banks village of Salvo, there's barely room left inside the post office for the postmaster. It's something Edward Hooper has always bragged about.

The post office in Salvo—Hooper's home away from home—measures eight feet by twelve feet. There's a tiny office in back crammed with not one but two desks. Out front, when three people walk in at the same time to check their boxes, it doesn't take a tape measure to figure out how small the lobby is.

The United States Postal Service took a survey of post offices based on size, local population, number of post-office boxes, and number of delivery-route stops. Salvo tied with post offices in Birds Landing, California, and Ochopee, Florida, as the smallest in the country.

The post office in Salvo was established in 1901. The building where Hooper works was constructed about twelve years later. Lafayette Douglass built it in his yard for his wife, Marcia, the postmaster. When Melvina Whidbee took over the position in 1947, she bought the building and moved it to her yard. Hooper bought it in 1977—he's reluctant to say how much he paid—and moved it on a boat trailer to his house. He plans to keep it there even after he retires.

"It's sentimental to me," he says. "I don't know what I'll do, but I'll keep it. They'll probably build another one, or I'll rent it to them."

Salvo is the kind of community where everyone knows the postmaster. But Hooper says the village has grown so fast he no longer knows everyone by name. No one knows for sure just how many residents there are, but there are ninety-four mailboxes.

Salvo sits on Hatteras Island a couple of miles south of the village of Waves. The town was originally named Clark. During the Civil War, when Union forces headed north after taking Hatteras Island, the commander of a ship asked what the village's name was. The naval chart didn't say. The commander ordered a salvo of cannon fire and entered the name *Salvo* on the chart. Salvo it became.

It's easy to miss. But once you're there, it's hard to miss the distinctive red, white, and blue post office—unless the mail truck has stopped to make a delivery. The truck is bigger than the post office.

EDWARD HOOPER PRIDES HIMSELF ON THE TINY SALVO POST OFFICE.

PHOTOGRAPH BY DEIDRA LAIRD

When Arnold And Miriam Daniels Of Wanchese Talk, You Can Hear Traces Of An English Accent.

Photograph by Deidra Laird

. . .

"Oi Love To Hear You Talk"

As Arnold Daniels spins a tale, his words rise and fall like waves on a calm summer evening.

He speaks in a thick brogue that harkens back several hundred years to when his ancestors first settled the fishing village of Wanchese, cut off from their homeland of England by the Atlantic Ocean and from the rest of the New World colonies by Croatan Sound.

"Thare's people in my toime never was off the oiland, only by bowt to go out fishin'. We were so oisolated, and we just talked just loike parents before us talked, you see. Naturally, it was mostly the people in this area we were around makin' contact with, so Oi guess that's the reason why we kept this brogue."

Daniels is known to outsiders—foreigners, he calls them—as a "Hoi Toider" because of the way he and other residents of the Outer Banks round the sound of the letter *i* in words like *high* and *tide*. It is a trace of a British accent carried over from seventeenth-century England and nurtured by years of isolation.

"The language simply developed on its own merry way," says linguist Bob Howren. "It's not Elizabethan English and they're not Irish. It's good old American English of a different stripe."

Scholars theorize that settlers from southwestern England carried the accent to coastal Virginia in the latter half of the seventeenth century. As they moved south to remote areas of the Outer Banks, they took the dialect with them. The British accent survived because of the area's remoteness.

Arnold Daniels's wife, Miriam, likes to tell of a trip to Dallas when a waitress asked her the perennial question, "Where are you from?"

Miriam Daniels answered that she was from England.

The waitress happened to be newly arrived in Texas from England, and she was eager to talk with a fellow Briton.

Daniels strung her on for a while before confessing, "Oi lied just a little bit. But Oi am English. And there's no doubt of it because Oi have poiple coming into the post office, where Oi've worked sixteen years, and they all said that an area around Devonshire—Oi could go there and they would think that was my hoome. . . . Oi told you a yarn because Oi'm so toired of hearing people make fun of the way Oi talk."

Wynne Dough, curator of the Outer Banks History Center in Manteo, says that "people living here tended to preserve patterns of speech lost elsewhere. 'Put some wind in that' is a Bankerism—something you don't hear anywhere else."

"Wind" means air, as in "Put some wind in the tire." Dough says that "chunk" means throw, as in "Chunk me the ball." "Mommick" means beat up, as in "The robber mommicked his victim." "Pizzer" means porch, as in "Come sit on the pizzer."

The Outer Banks region isn't unique. Distinctive British-sounding accents survive in parts of coastal Virginia, along the Maine coast, and in remote areas of the North Carolina mountains. The accents survive, Bob Howren says, in places "where people for generations have lived in relatively closed communities."

David Shores, an English professor at Old Dominion University, says, "The remarkable thing about these extreme outermost communities and dialects . . . is that they are closer to each other along the Atlantic seaboard than to the states for which they belong."

That's because it was easier to sail from one coastal community to another than it was to sail to the mainland. And since many islanders fished for a living, they had more reasons to travel the coast than to venture inland. It stayed that way along the North Carolina coast well into the twentieth century.

The first bridge connecting part of the Outer Banks to the mainland wasn't finished until the late 1930s. The three-mile bridge spanned Currituck Sound, connecting Bodie Island, where Nags Head and Kitty Hawk are located, with Point Harbor. The Croatan Sound Bridge, connecting Roanoke Island with the mainland, wasn't opened until the spring of 1957. Arnold Daniels remembers that day well. Until then, he and other islanders rode a five-hour ferry from Wanchese to Elizabeth City to get to the mainland, whether they needed to go to the hospital or shop in a department store.

With bridges and the increasing popularity of the Outer Banks as a resort, the local dialect has been contaminated by outside influences. Carole Hines, a linguist at Old Dominion University, noticed changes while she was studying the way Wanchese women speak. "When you move to later generations," she says, "you find that the children, for instance—they've had so many different influences, it's begun to change."

Some islanders who leave the Outer Banks lose their accent. Others return with a stronger accent. "It may be a way of asserting one's uniqueness," Wynne Dough says.

Outer Banks native Joe T. Daniels transported oil in Pennsylvania for many years before returning home to Wanchese in 1986. He kept his accent. In fact, it's stronger than that of his uncle, Arnold Daniels. "The other day Oi had to go to Philadelphia," Joe Daniels says. "The lawyer asked me to talk. He told me, 'Oi love to hear you talk.'"

JOE T. DANIELS HAS A STRONGER ACCENT THAN HIS GRANDSON STEPHEN "TEE" DANIELS.

PHOTOGRAPH BY DEIDRA LAIRD

Josephine Spencer Once Picked
Seventy-Five Pounds Of Crabmeat A Day.

Photograph by Deidra Laird

. . .

THE CRAB PICKER

JOSEPHINE SPENCER'S calloused hands fly across the table. She jerks the claws off a cooked crab, rips off the bright orange back, scoops out the eggs, cuts off the legs, then carefully picks out the delicate white meat. "I love to pick crabs," Spencer says, her bright brown eyes never even looking down to see what her hands are doing. Less than a minute and she's on to the next crab.

Spencer is a crab-picking expert. She's been making a living at it for more than thirty-five years. She says that while people don't get rich in her line of work, "you don't stay broke once you learn how to pick 'em."

She picks crabs in Wanchese, an out-of-the-way corner of Roanoke Island. "They put a timer on me one time long ago," she brags. "I used to be able to pick seventy-five pounds a day. I slowed down since I got older. Everything I do is fast. Everybody says that, even about my housework."

Spencer grew up in Engelhard on Pamlico Sound, where most of the state's forty crab houses are located. Crabs are big business on the North Carolina coast. Fishermen caught more than 31.8 million pounds in 1988, valued on the docks at more than $7 million. It's such a good business in the area that a local college hired Spencer to train more workers. Her job comes with a title: Crab-picking Instructor of the College of the Albemarle.

Spencer began picking at age seventeen, the year after she married. She worked in crab houses in Engelhard and nearby Mattamuskeet before the college sent her to St. Elmo's Crab Company in Wanchese to teach others to pick crabs.

In an age of automation, there's still only one way to pick crabs, and that's by hand. There's an art to it. The toughest part is making sure to pick only the meat, not the bones and other debris that often make their way into containers. "If the crab is cut deep enough," Spencer says, "you always get good clean meat. That's what I tell them. I check out every pound. Oh my gracious, I like to work with the meat. I like to place the meat, fix it up. . . . Not that I couldn't quit. I'm an ordained minister."

Spencer always works with the same four-inch stainless-steel knife. No other knife will do. She sits on a chair in front of a long stainless-steel table. At five-foot-three, she's too short to reach the floor of the concrete-block building, so she rests her feet on a spare block.

Nineteen-year-old Lisa Barnett works across the table. Her face contorts in frustration as she slowly picks at a crab. There's none of Spencer's speed. There's none of the joy Spencer radiates when she's finished picking, when she flashes a smile so broad it reveals two gold teeth.

"I don't like this," Barnett admits. "I don't know, maybe I could, maybe I couldn't. I can only do ten pounds a day."

"Oh my gracious," Spencer says. "You can do more than ten pounds, Lisa."

"I can. But I don't."

Patty Jarvis, co-owner of Mattamuskeet Seafood, south of Wanchese, says she expects each worker to pick twenty pounds in an eight-hour day. Most pickers get paid by the pound, at the

THERE'S STILL ONLY ONE WAY TO PICK CRABS—BY HAND.

PHOTOGRAPH BY DEIDRA LAIRD

FREDDIE BARNES SHOVELS ANOTHER LOAD OF CRABS FOR PICKING.

PHOTOGRAPH BY DEIDRA LAIRD

going rate of $1.25 to $1.50 per pound. If they're good, the arrangement generally works out better than minimum wage. But Donald Caroon, the owner of St. Elmo's Crab Company, where Spencer teaches, pays his trainees by the hour instead of by the pound. They go through about two thousand pounds of crabmeat a day, compared with ten thousand pounds at Mattamuskeet Seafood.

"It's slow going," Spencer says of her trainees. "But they'll catch on."

Freddie Barnes shovels a load of crabs onto the table. Then he stares at Spencer. "I just like to sit here and watch her," he says. "She can put on a good show. She can pick and do a whole lot of talking." And singing, too. Spencer is as likely to break into a rendition of "God Has Smiled on Me" as she is to scold another worker that "the Bible says he who has no sin casts the first stone."

By the end of the day, she's as aromatic as the crabs she picks. "Your whole skin smells," she says. "You really got to take a good bath. My husband, he can't stand the smell. Now, crab picking has become a part of me. I don't eat many. I been working in seafood so long crabs isn't one of my favorites. I just like to pick 'em."